高等职业院校电力技术类专业系列教材

电气设备检修（二）
——变压器类（中英双语）

主　编　华　章　邓　浩
副主编　刘　硕　程　铭　陈　丽
参　编　杨　熙　刘　燕　张煌竟　陈　杰　姜聿涵　张振东

西南交通大学出版社
·成　都·

图书在版编目（CIP）数据

电气设备检修. 二，变压器类：汉、英／华章，邓浩主编. --成都：西南交通大学出版社，2023.11
ISBN 978-7-5643-9583-4

Ⅰ. ①电… Ⅱ. ①华… ②邓… Ⅲ. ①变压器 – 设备检修 – 高等职业教育 – 教材 – 汉、英 Ⅳ. ①TM64

中国国家版本馆 CIP 数据核字（2023）第 229770 号

Dianqi Shebei Jianxiu (Er)
—Bianyaqi Lei (Zhong-Ying Shuangyu)

电气设备检修（二）——变压器类（中英双语）	主编　华　章　邓　浩	策划编辑　李芳芳　张少华 责任编辑　孟媛 封面设计　吴兵

印张：17.75　　字数：481千	出版发行　西南交通大学出版社
成品尺寸：185 mm×260 mm	网址　http://www.xnjdcbs.com
版次：2023年11月第1版	地址　四川省成都市二环路北一段111号 　　　西南交通大学创新大厦21楼
印次：2023年11月第1次	邮政编码　610031
印刷　四川玖艺呈现印刷有限公司	营销部电话　028-87600564　028-87600533
书号　ISBN 978-7-5643-9583-4	定价：65.00元

图书如有印装质量问题　本社负责退换
版权所有　盗版必究　举报电话　028-87600562

FOREWORD

为贯彻《国务院关于加快发展现代职业教育的决定》文件精神，更好地满足高等职业教育高质量发展的需要，实现教学内容由知识本位向能力本位的转变，结合人力资源和社会保障部国家职业技能标准《变电设备检修工》的职业能力要求，以提高电气设备检修人员的技术技能水平为目标，为满足市场和企业不断发展的岗位需求，编写了本套教材。本套教材响应"一带一路"建设号召，服务国家战略方针，深化电力行业"一带一路"交流合作。本套教材具备代表性，基础性，针对性及普遍性等特点。本教材的出版有利于拓展国际业务交流，加强电力文化输出，提升国家影响力。

本套教材共分为3本，利用"校企一体，师资互用"机制，践行"双师"结构与"双师"资质，与生产企业共同编制，由专业教师和企业教师一同进行课程内容深度分析。本套教材的编制，旨在以学生为中心，落实立德树人的根本任务，在使学生掌握相关职业资格技能鉴定或教育部"1+X"变电设备检修职业技能等级考试标准制度试点职业技能等级所需的知识和技术技能，满足相关工种的中级工或以上要求的同时，培养学生吃苦耐劳、团结协作、可迁移可转化、注重安全等职业素养和行为习惯，弘扬工匠精神，达到"德技并修、理实并重、手脑并用、讲赛并行、工学结合"的要求。

本书是电气设备检修课程的核心组成部分，鉴于在实际工作中的变压器，特别是110 kV及以上的主变不可能随着学员的学习需求随时吊罩、解体，也不可能随时停电去检修带电附件，因此本书除了检修工作必须要了解的理论知识外，归纳和整理了一些具有典型代表性的变压器的本体和器身检修操作工作，旨在通过由外及内的变压器检修对象，读者能逐步深入了解变压器结构与工作原理的产品应用和表现形式，熟悉变压器运行维护、检修及故障处理的操作要求和质量标准，强调相关国家、行业、企业的规程、规范和标准的引用和落实，突出技能训练，注重培养学员实际动手操作能力，尽量满足变压器设备制造、安装、运行、检修、销售技术服务等生产岗位零距离上岗的要求。

本书根据理实一体化教学的需要，以项目导向、任务驱动为主线，学习内容遵循由浅入深、循序渐进的原则，采用教室+实训现场的理论实践相结合的教学方法，充分体现了教、学、做一体化。本书在编写过程中突出"工作任务导向、规范作业流程、理论知识够用，突出技能实训"的思想，强调安全作业和标准化作业。全书实操内容较为典型，按教学项目和教学模块设计，突出技能训练，使学员通过对技术工作的任务、过程、环境所进行的整体化感悟和反思，实现知识与技能、过程与方法、情感态度和价值观学习的统一。

本书依托国网四川省电力公司技能培训中心（四川电力职业技术学院）丰富的电气设备检修实训资源，采用了大量源自生产作业现场和技能实训现场的实拍图片，增强了本书的实用性和可读性。

本书由国网四川省电力公司技能培训中心（四川电力职业技术学院）华章、国网四川省德阳供电公司高级技师邓浩担任主编，刘硕、程铭、陈丽担任副主编。全书编写分工如下：项目一电气设备概述由杨熙、陈杰、邓浩编写，项目二变压器检修由华章、张振东、姜聿涵编写，项目三互感器检修由刘燕、刘硕、程铭编写，项目四其他变电设备检修由张煌竟、陈丽编写，华章负责全书内容的审定。

本书由国网四川省电力公司乐山供电公司高级技师李运涛、攀枝花供电公司高级技师唐启刚、遂宁供电公司高级技师赵安主审。编写过程中得到国网四川省电力公司技能培训中心（四川电力职业技术学院）汤晓青副教授和电网检修部（电力设备技术系）同事的大力支持，在此表示衷心的感谢！

限于编者水平，书中不足和错误之处在所难免，恳请读者批评指正，不胜感激。

编　者

2023 年 5 月

CONTENTS

项目一　电气设备概述 ···001
　　模块一　发电厂及变电站概述 ···001
　　Module 1　Overview of Power Plants and Substations ·······························013
　　模块二　主要电气设备 ···028
　　Module 2　Main Electrical Equipment ··030
　　模块三　电气设备符号及装置概述 ···034
　　Module 3　Overview of Electrical Equipment Symbols and Devices ············036

项目二　变压器检修 ···039
　　模块一　变压器检修周期 ··039
　　Module 1　Transformer Maintenance Cycle ··041
　　模块二　变压器部件和基础知识 ··043
　　Module 2　Transformer Components and Basics ···································049
　　模块三　变压器绝缘 ··056
　　Module 3　Transformer Insulation ··061
　　模块四　变压器部件功能 ··066
　　Module 4　Transformer Component Functions ·····································077
　　模块五　变压器专业巡视内容和评价标准 ···093
　　Module 5　Content of Specialized Inspections and Evaluation
　　　　　　　Criteria for Transformers ···096
　　　任务一　变压器吊罩后器身检查 ···099
　　　Task 1　Transformer Body Inspection after Core Hoisting ··················109
　　　任务二　M 型有载分接开关机构箱检修 ···124
　　　Task 2　Maintenance of Mechanism Box of M-Type On-Load
　　　　　　　Tap Changer ···143
　　　任务三　变压器呼吸器（吸湿器）小修 ··174
　　　Task 3　Minor Repair of Transformer Breather (Moisture Absorber) ······177

项目三　互感器检修 ·· 181

模块一　互感器概述 ··· 181
Module 1　Transformer Overview ··································· 182
模块二　电压互感器 ··· 183
Module 2　Voltage Transformer ····································· 193
模块三　电流互感器 ··· 205
Module 3　Current Transformer ····································· 212

任务一　互感器极性的测试方法直流法判断互感器的极性 ············ 221
Task 1　Transformer Polarity Testing Method Determination of Transformer Polarity with DC Method ···················· 225
任务二　电流互感器的变比改接 ··· 231
Task 2　Wiring Change Based on Ratio of Transformation of Current Transformer ·· 235

项目四　其他变电设备检修 ·· 240

模块一　母线 ··· 240
Module 1　Bus ·· 244
模块二　绝缘子 ·· 248
Module 2　Insulator ·· 253
模块三　高压熔断器 ·· 260
Module 3　High-voltage Fuse ·· 265
模块四　补偿设备 ··· 271
Module 4　Compensator ··· 273

参考文献 ·· 277

项目一 电气设备概述

模块一 发电厂及变电站概述

一、发电厂

发电厂是把各种天然能源（化学能、水能、原子能等）转换成电能的工厂。按使用能源不同或转换能源特点，发电厂有以下类型。

（一）火力发电厂

火力发电厂是把化石燃料（煤、油、天然气、油页岩等）的化学能转换成电能的工厂，简称火电厂。火电厂的原动机大都为汽轮机，也有燃气轮机、柴油机等。火电厂又可分为以下几种。

1. 凝汽式火电厂

凝汽式火力发电厂的生产过程如图 1-1 所示。煤粉在锅炉炉膛中燃烧，使锅炉中的水加热变成过热蒸汽，经管道送到汽轮机，推动汽轮机旋转，将热能变为机械能。汽轮机带动发电机旋转，再将机械能变为电能。在汽轮机中做过功的蒸汽排入凝汽器，循环水泵打入的循环水将排汽迅速冷却而凝结，由凝结水泵将凝结水送到除氧器中除氧（清除水中的气体，特别是氧气），而后由给水泵重新送回锅炉。

由于在凝汽器中大量的热量被循环水带走，因此，凝汽式火电厂的效率较低，只有 30%~40%。

2. 热电厂

热电厂生产过程的如图 1-2 所示。热电厂与凝汽式火电厂不同之处是将汽轮机中一部分做过功的蒸汽从中段抽出来直接供给热用户，或经热交换器将水加热后，把热水供给用户。这样，便可减少被循环水带走的热量，提高效率，现代热电厂的效率一般为 60%~70%。

由于供热网络不能太长，所以热电厂总是建在热力用户附近。此外，为了使热电厂维持较高的效率，一般采用"以热定电"的运行方式，即当热力负荷增加时，热电机组相应地多发电，当热力负荷减少时，热电机组相应地少发电。因而，其运行方式不如凝汽式发电厂灵活。

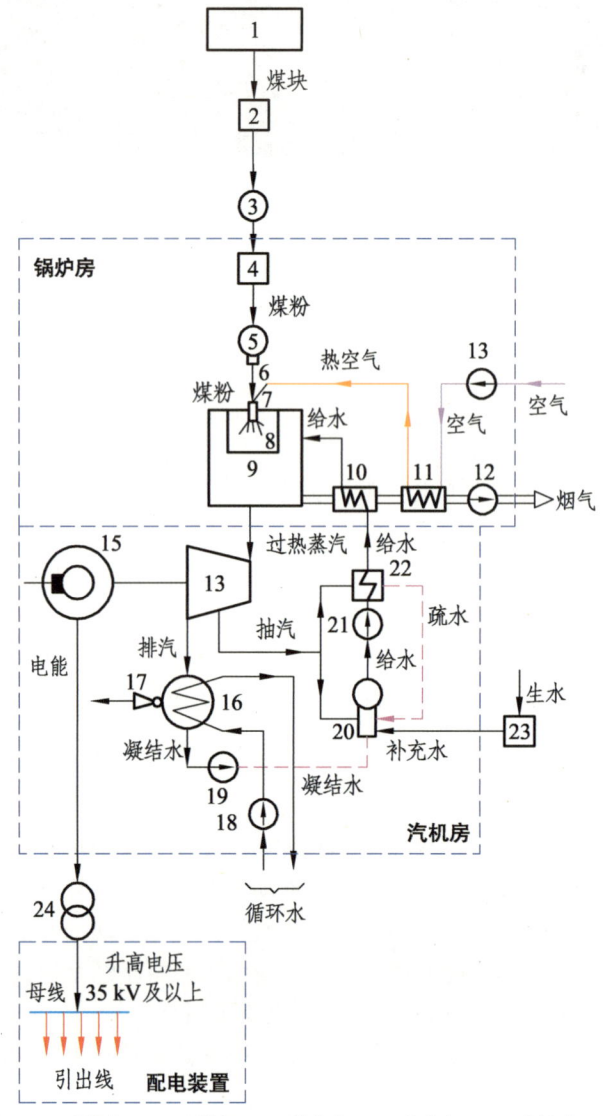

1—煤场；2—碎煤机；3—原煤仓；4—磨煤机；5—煤粉仓；6—给粉机；7—喷燃器；8—炉膛；9—锅炉；
10—省煤器；11—空气预热器；12—引风机；13—送风机；14—汽轮机；15—发电机；16—凝汽器；
17—抽气器；18—循环水系；19—凝结水泵；20—除氧器；21—给水泵；
22—加热器；23—水处理设备；24—升压变压器。

图 1-1　凝汽式火电厂生产过程的示意图

3. 燃气轮机发电厂

用燃气轮机或燃气-蒸汽联合循环中的燃气轮机和汽轮机驱动发电机的发电厂，称为燃气轮机发电厂。前者一般用作电力系统的调峰机组，后者则用来带中间负荷和基本负荷。这类发电厂可燃用液体燃料或气体燃料。以天然气为燃料的燃气轮机和联合循环发电，具有效率高、污染物排放低、初期投资少、工期短及易于调节负荷等优点，近年来在北美、欧洲得到迅速发展。目前燃气轮机的单机容量已发展到30万千瓦。

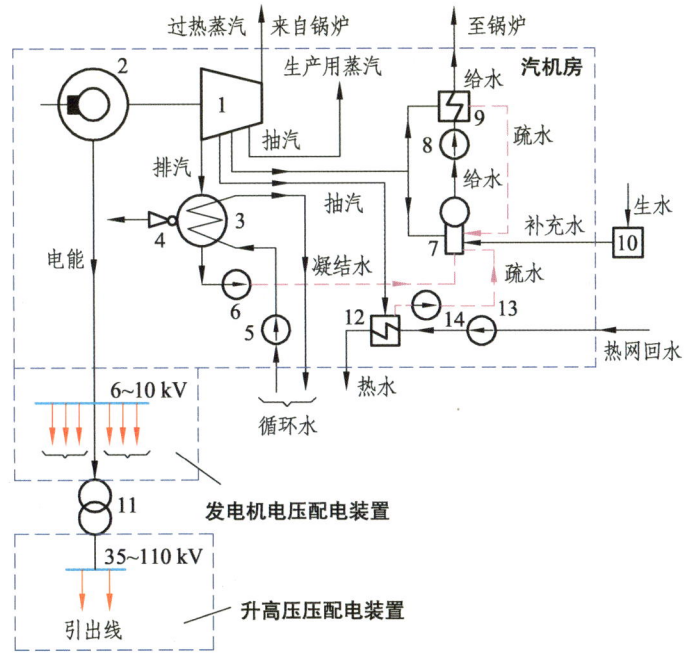

1—汽轮机；2—发电机；3—凝汽器；4—抽气器；5—循环水泵；6—凝结水泵；7—除氧器；8—给水泵；
9—加热器；10—水处理设备；11—升压变压器；12—加热器；13—回水泵；14—泵。

图 1-2　热电厂生产过程的示意图

燃气轮机的工作原理与汽轮机相似，不同的是其工质不是蒸汽，而是高温高压气体。空气经压气机压缩增压后送入燃烧室，燃料经燃料泵打入燃烧室。燃烧产生的高温高压气体进入燃气轮机中膨胀做功，推动燃气轮机旋转，带动发电。做过功的的尾气经烟囱排出，或分流部分用于制热、制冷。这种单纯用燃气轮机驱动发电机的发电厂，热效率只有 35%～40%。

为提高热效率，采用燃气-蒸汽联合循环系统，图 1-3 是其模式之一。燃气轮机的

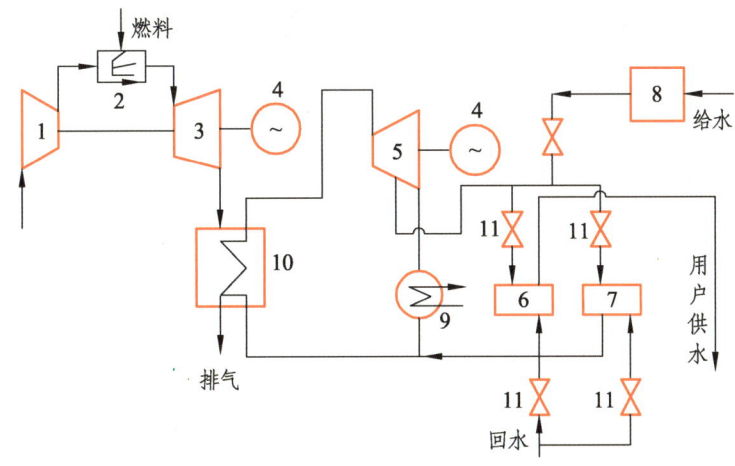

1—压气机；2—燃烧室；3—燃气轮机；4—发电机；5—汽轮机；6—蒸汽型溴冷机；7—汽-水热交换器；
8—备用燃气锅炉；9—凝汽器；10—余热锅炉；11—制冷采暖切换阀。

图 1-3　燃气—蒸汽联合循环系统

排气进入余热锅炉，加热其中的给水并产生高温高压蒸汽，送到汽轮机中去做功，带动发电机再次发电；从汽轮机中抽取低压蒸汽（发电机停止发电时起动备用燃气锅炉提供汽源），通过蒸汽型溴冷机（溴化锂作为吸收剂）或汽-热交换器制取冷、热水。这是电、热、冷三联供模式。联合循环系统的热效率为 56%～85%。

（二）水力发电厂

水力发电厂是把水的位能和动能转换成电能的工厂，简称水电厂，也称水电站。水电站的原动机为水轮机，通过水轮机将水能转换为机械能，再由水轮机带动发电机将机械能转换为电能。

1. 坝式水电站

在河流上的适当地方建筑拦河坝，形成水库，抬高上游水位，使坝的上、下游形成大的水位差，这种水电站称为坝式水电站。坝式水电站适宜建在河道坡降较缓且流量较大的河段。这类水电站按厂房与坝的相对位置又可为以下几种。

（1）坝后式厂房。坝后式水电站如图 1-4 所示。其厂房建在拦河坝非溢流坝段的后面（下游侧），不承受上游水的压力，压力管道通过坝体，适用于高、中水头。

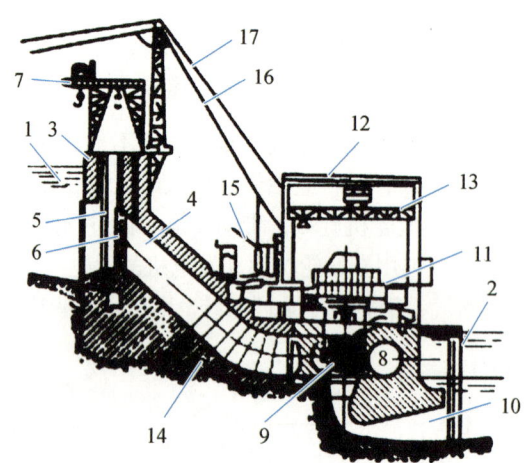

1—上游水位；2—下流水位；3—坝；4—压力进水管；5—检修闸门；6—闸门；7—吊车；8—水轮机蜗壳；
9—水轮机转子；10—尾水管；11—发电机；12—发电机间；13—吊车；14—发电机电压配电装置；
15—升压变压器；16—架空线；17—避雷线。

图 1-4 坝后式水电站断面图

水电站的生产过程较简单，发电机与水轮机转子同轴连接，水由上游沿压力水管进入水轮机蜗壳，冲动水轮机转子，水轮机带动发电机转动即发出电能；做过功的水通过尾水管流到下游；生产出来的电能经变压器升压并沿架空线至屋外配电装置，而后送入电力系统。

（2）溢流式厂房。溢流式厂房建在溢流坝段后（下游侧），泄洪水流从厂房顶部越过泄入下游河道，适用于河谷狭窄，水库下泄洪水流量大，溢洪与发电分区布置有一定困难的情况。

（3）岸边式厂房。岸边式厂房建在拦河坝下游河岸边的地面上，引水道及压力管道明铺于地面或埋没于地下。

（4）地下式厂房。地下式厂房的引水道和厂房都建在坝侧地下。

（5）坝内式厂房。坝内式厂房的压力管道和厂房都建在混凝土坝的空腔内，且常设在溢流坝段内，适用于河谷狭窄，下泄洪水流量大的情况。

（6）河床式厂房。河床式厂房的水电站如图 1-5 所示。其厂房与拦河坝相连接，成为坝的一部分，厂房承受水的压力，适用于水头小于 50 m 的水电站。图中的溢洪坝、溢洪道是为了渲泄洪水、保证大坝安全的泄水建筑物。

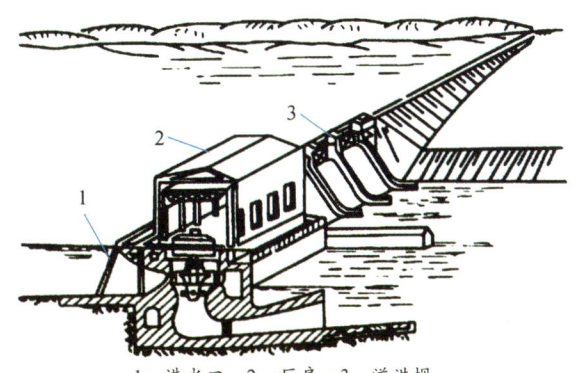

1—进水口；2—厂房；3—溢洪坝。

图 1-5　河床式水电站示意图

2. 引水式水电站

由引水系统将天然河道的落差集中进行发电的水电站，称为引水式水电站。引水式水电站适宜建在河道多弯曲或河道坡降较陡的河段，用较短的引水系统可集中较大的水头；也适用于高水头水电站，避免建设过高的挡水建筑物。

引水式水电站如图 1-6 所示。在河流适当地段建低堰（挡水低坝），水经引水渠和压力水管引入厂房，从而获得较大的水位差。

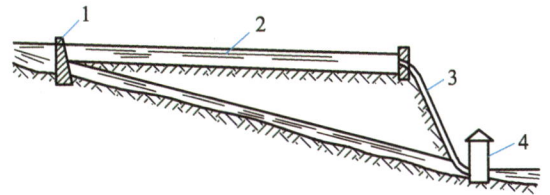

1—挡水低坝；2—引水渠；3—压力水管；4—厂房。

图 1-6　引水式水电站

3. 抽水蓄能电站

利用电力系统低谷负荷时的剩余电力抽水到高处蓄存，在高峰负荷时放水发电的水电站，称为抽水蓄能电站。它是电力系统的填谷调峰电源。在以火电、核电为主的电力系统中，建设适当比例的抽水蓄能电站可以提高系统运行的经济性和可靠性。抽水蓄能电站可能是堤坝式或引水式。

抽水蓄能电站如图 1-7 所示。当电力系统处于低谷负荷时，其机组以电动机-泵方式工作，吸收电力系统的有功功率将下游的水抽至上游水库蓄存起来，把电能转换为水能，这时它是用户；当电力系统处于高峰负荷时，其机组按水轮机-电机方式运行，使所蓄的水用于发电，以满足调峰需要，这时它是发电站。

（三）核电厂

核电厂是将原子核的裂变能转换为电能的发电厂，燃料主要是 U235。U235 容易在慢中子的撞击下裂变，释放出巨大能量，同时释放出新的中子。按所使用的慢化剂和冷却剂，核反应堆可分为轻水堆、重水堆、石墨气冷堆及石墨沸水堆。其中轻水堆又分压水堆和沸水堆。

核电厂的生产过程与一般火电厂相似，即将核能产生的热能，再按火电厂的发电方式将热能转换为机械能，再转换为电能。核电厂中以轻水堆（压水堆和沸水堆）核电厂最多。轻水堆式核电厂发电方式示意图如图 1-8 所示。

1—输水系统；2—厂房和水库；3—上水库

图 1-7 抽水蓄能电站

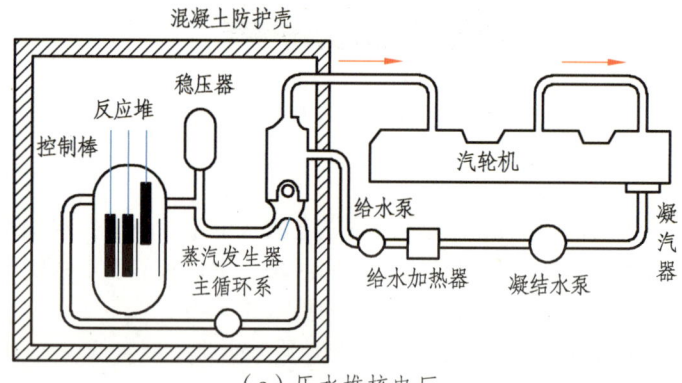

（a）压水堆核电厂

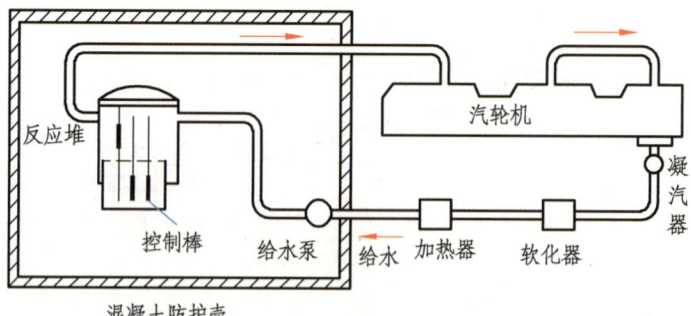

（b）沸水堆核电厂

图 1-8 轻水堆式核电厂发电方式示意图

压水堆核电厂实际上是用核反应堆和蒸汽发生器代替一般火电厂的锅炉。反应堆中通常有 100~200 个燃料组件。在主循环水泵（又称压水堆冷却剂泵或主泵）的作用下，压力为 15.2～15.5 MPa、温度 290 ℃左右的蒸馏水不断在左回路（称一回路，有 2～4 条并联环路）中循环，经反应堆时被加热到 320 ℃左右，然后进入蒸汽发生器，并将自身的热量传给右回路（称二回路）的给水，使之变成饱和或微过热蒸汽；蒸汽沿管道进入汽轮机膨胀做功，推动汽轮机并带动发电机发电。二回路的工作过程与火电厂相似。

压水堆的快速变化反应性控制，主要是通过改变控制棒（内装银–铟–镉材料的中子吸收体）在堆芯中的位置来实现。

左回路中稳压器（带有安全阀和卸压阀）的作用是在电厂启动时用于系统升压（力），在正常运行时用于自动调节系统压力和水位，并提供超压保护。

沸水堆核电厂是以沸腾轻水为慢水剂和冷却剂并在反应堆内直接产生饱和蒸汽，通入汽轮机做功发电；汽轮机的排汽冷凝后，经轻化器净化、加热器加再由给水泵送入反应堆。

（四）新能源发电

1. 风力发电

流动空气所具有的能量，称为风能。全球可利用的风能约为 2×10^6 万千瓦。风能属于可再生能源，是一种过程性能源，不能直接储存，而且具有随机性，这给风能的利用增加了技术上的复杂性。

将风能转换为电能的发电方式，称为风力发电。在风能丰富的地区，按一定的排列方式成群安装风力发电机组，组成集群，称为风力发电场。其机组多达几十台、几百台，甚至数千台，是大规模开发利用风能的有效形式。

风力发电装置如图 1-9 所示。风力机将风能转化为机械能（属于低速旋转机械），升速齿轮箱将风力机轴上的低速旋转变为高速旋转，带动发电机发出电能；经电缆线路引至配电装置，然后送入电网。

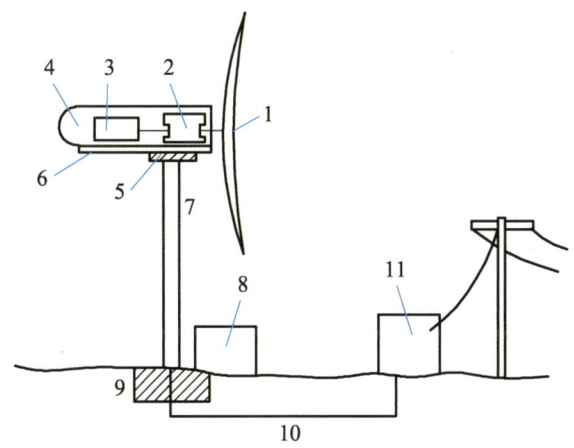

1—风力机；2—升速齿轮箱；3—发电机；4—控制系统；5—驱动装置；6—底板和平外罩；7—塔架；
8—控制和保护装置；9—土建基础；10—电缆线路；11—配电装置。

图 1-9　风力发电装置

风力机的叶片（2~3叶）多数是由聚酯树脂增强玻璃纤维材料制成，塔架由钢材制成（锥形筒状式或桁架式），升速齿轮箱一般为三级齿轮传动。风力发电机组的单机容量为几十瓦至几兆瓦，100kW以上的风力发电机为同步发电机或异步发电机。大、中型风力发电机组皆配有由微机或可编程控制器组成的控制系统，以实现控制、自检、显示等功能。

2. 海洋能发电

海洋能是蕴藏在海水中的可再生能源，如潮汐能、波浪能、海流能、海洋温差能、海洋盐差能等。潮汐发电就是利用潮汐的位能发电，即在潮差大的海湾入口或河口筑堤构成水库，在坝内或坝侧安装水轮发电机组，利用堤坝两侧的潮差驱动水轮发电机组发电。

（1）单库单向式。单库单向式潮汐发电站如图1-10所示。这种发电站只建一个水库，安装单向水轮发电机组（发电机安装于密封的灯泡体内），在落潮时发电。当涨潮至库内水位时，开闸向水库充水，至库内外在更高的水位齐平时关闸，等待潮水逐渐下降；当库内外水位差达机组启动水头时开闸发电（这时水库水位逐渐下降），直到库内外水位差小于机组发电所需的最低水头，再次关闸等待，转入下一周期。

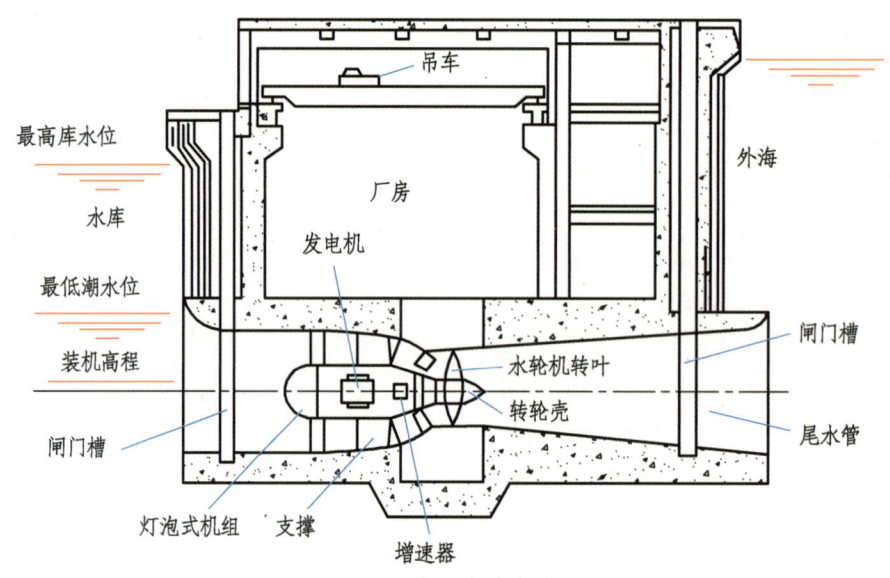

图1-10 单库单向式潮汐电站

（2）单库双向式。单库双向式潮汐发电站如图1-11所示。这种发电站也只建一个水库，安装双向水轮发电机组，在涨落潮时均发电。当涨潮到一定高度时，打开闸A、B将潮水引入站内冲动机组发电；当涨潮将结束时，迅速打开闸E、F，使水库充满水后即关闸；当落潮至一定水位差时，打开闸C、D再次冲动机组发电。这样实现了涨落潮双向发电。

（3）双库（高低库）式。建两个毗连的水库，水轮发电机组安装在两水库之间的隔坝内。高库设有进水闸，在潮位较库内水位高时进水（低库不进水），以尽量保持高水位；低库设有泄水闸，在潮位较库内水位低时泄水。这样，两库之间终日有水位差，可连续发电。

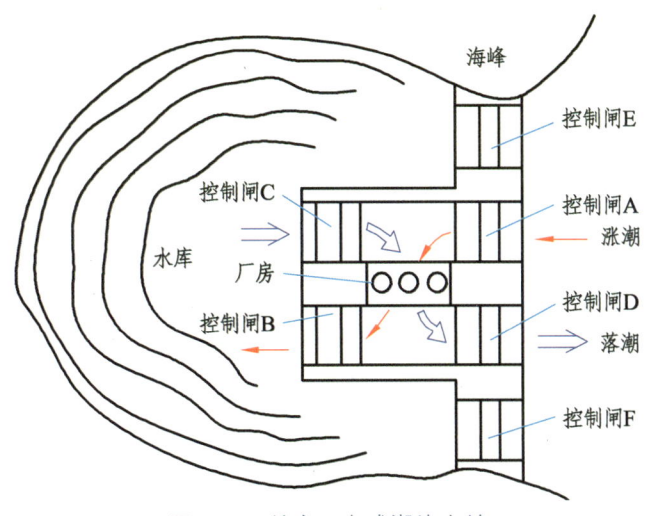

图 1-11 单库双向式潮汐电站

3. 地热发电

利用地下蒸汽或热水等地球内部热能资源发电,称为地热发电。目前,热发电的单机容量最大为 15 万千瓦。地热蒸汽发电的原理和设备与火电厂基本相同。利用地下热水发电,有两种基本类型。

(1) 闪蒸地热发电系统 (又称减压扩容法)。此方法是使地下热水变为低压蒸汽供汽轮机做功,如图 1-12 所示。地下热水经除氧器除氧后,进入第一级扩容器进行减压扩容,产生一次蒸汽 (约占热水量的 10%),送入汽轮机的高压部分做功,余下的热水进入第二级扩容器,再进行二次减压扩容,产生二次蒸汽,因其压力低于第一级,所以送入汽轮机的低压部分做功。实际采用的扩容级数一般不超过四级。我国羊八井地热电站为两级扩容。

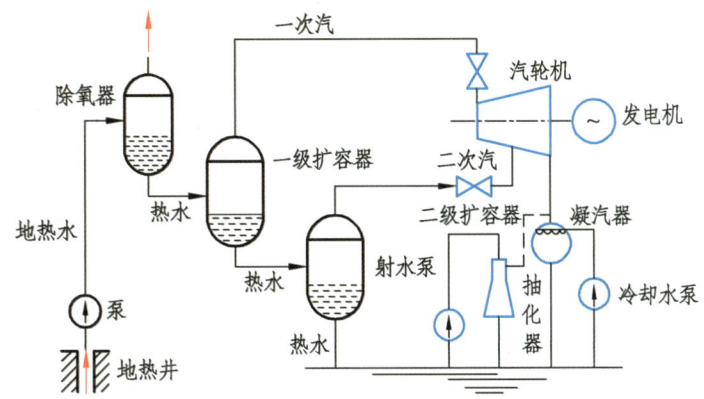

图 1-12 闪蒸地热发电系统

扩容蒸发又称闪蒸。当将具有一定压力及温度的地热水注入到压力较低的容器中,由于水温高于容器压力的饱和温度,一部分热水急速汽化为蒸汽,并使温度降低,直到水和蒸汽都达到该压力下的饱和状态为止。当地热井口流体为湿蒸汽时,则先进入汽水分离器,分离出的蒸汽送往汽轮机,剩余的水再进入扩容器。

（2）双循环地热发电系统（又称中间介质法）。其流程如图 1-13 所示。地下热水用深井泵抽到电站的蒸发器内，加热某种低沸点工质（如氟利昂、异丁烷、正丁烷等），使其变成低沸点工质蒸汽，推动汽轮发电机发电；汽轮机的排汽经凝汽器冷凝成液体，用工质泵再打回蒸发器重新加热循环使用。为充分利用地热水的余热，从蒸发器排出的地热水经预热器先预热来自凝汽器的低沸点工质液体。这种系统的热水和工质各自构成独立系统，故称双循环系统。

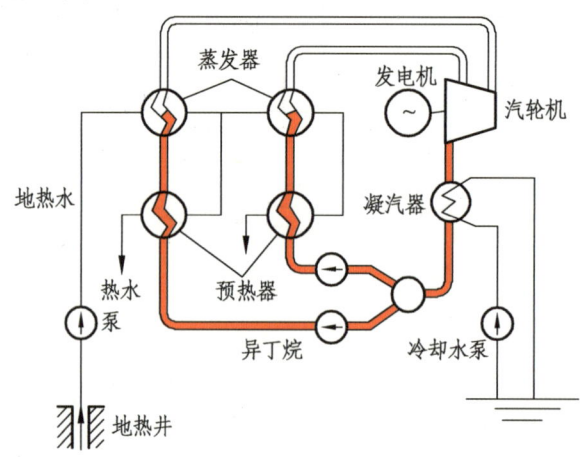

图 1-13 双循环地热发

4. 太阳能发电

太阳能是从太阳向宇宙空间发射的电磁辐射能，到达地球表面的太阳能为 8.2×10^9 万千瓦，能量密度为 $1 kW/m^2$ 左右。太阳能发电有热发电和光发电两种方式。

（1）太阳能热发电。太阳能热发电是将吸收的太阳辐射热能转换成电能的装置，其基本组成与常规火电设备类似。它又分为集中式和分散式两类。

集中式太阳能热发电又称塔式太阳能热发电，其热力系统流程如图 1-14 所示。它是在很大面积的场地上整齐地布设大量的定日镜（反射镜）阵列，且每台都配有跟踪系统，准确地将太阳光反射集中到一个高塔顶部的吸热器（又称接收器）上，把吸收的光能转换成热能，使吸热器内的工质（水）变成蒸汽，经管道送到汽轮机，驱动机组发电。

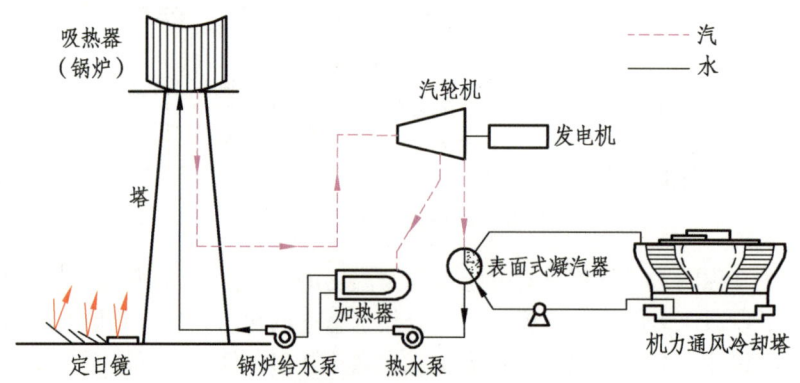

图 1-14 塔式太阳能电站热力系统流程

分散式太阳能热发电，是在大面积的场地上安装许多套结构相同的小型太阳能集热装置，通过管道将各套装置所产生的热能汇集起来，进行热电转换，发出电力。

（2）太阳能光发电。太阳能光发电是不通过热过程而直接将太阳的光能转变成电能，有多种发电方式，其中光伏发电方式是主流。光伏发电是把照射到太阳能电池（也称光伏电池，是一种半导体器件，受光照射会产生伏打效应）上的光直接变换成电能输出。

5. 生物质能发电

生物质能是绿色植物通过叶绿素将太阳能转化为化学能而储存在生物质内部的能量，属可再生能源。薪柴、农作物秸秆、人畜粪便、有机垃圾及工业有机废水等，是主要的生物质能资源。生物质发电系统是以生物质能为能源的发电工程，如垃圾焚烧发电、沼气发电、蔗渣发电等。

6. 磁流体发电

磁流体发电亦称等离子体发电，是使极高温度并高度电离的气体高速（1000 m/s）流经强磁场而直接发电。这时气体中的电子受磁力作用和气体中活化金属粒子（钾、铯）相互碰撞，沿着与磁力线成垂直的方位流向电极而发出直流电。

二、变电站

变电站是联系发电厂和用户的中间环节，起着变换和分配电能的作用。变电站有多种分类方法，可以根据电压等级、升压或降压及在电力系统中的地位分类。根据变电站在系统中的地位，可分为以下几类。各类变电站的示意图如图 1-15 所示。

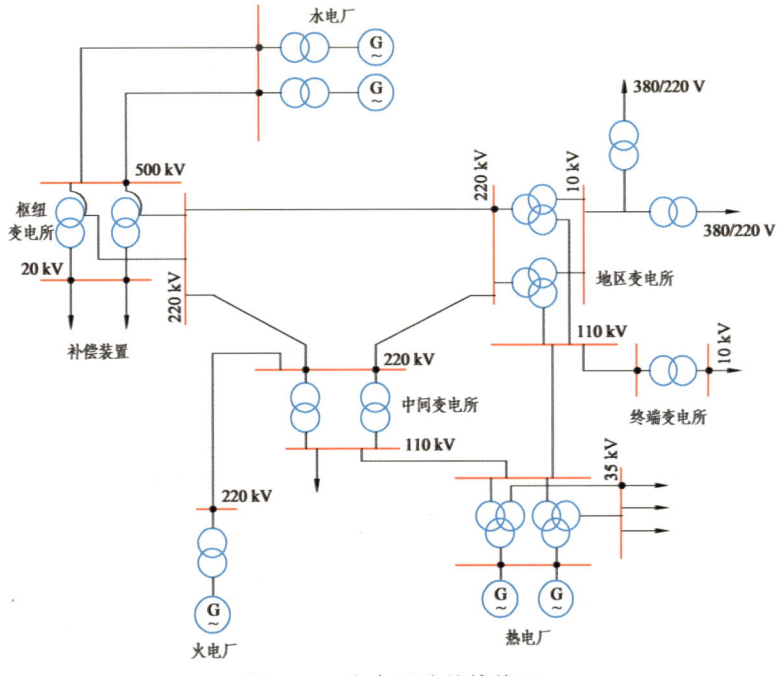

图 1-15 电力系统的接线图

1. 枢纽变电站

枢纽变电站位于电力系统的枢纽点，连接电力系统高、中压的几个部分，汇集有多个电源和多回大容量联络线，变电容量大，电压（指其高压侧，以下同）等级为 330～500 kV。全站停电时，将引起系统解列，甚至瘫痪。

2. 中间变电站

中间变电站一般位于系统的主要环路线路中或系统主要干线的接口处，汇集有 2～3 个电源，高压侧以交换潮流为主，同时又降压供给当地用户，主要起中间环节作用，电压等级为 220～330 kV。全站停电时，将引起区域电网解列。

3. 地区变电站

地区变电站以对地区用户供电为主，是一个地区或城市的主要变电站，电压等级一般为 110～220 kV。全站停电时，仅使该地区中断供电。

4. 终端变电站

终端变电站位于输电线路终端，接近负荷点，经降压后直接向用户供电，不承担功率转送任务，电压等级为 110 kV 及以下。全站停电时，仅使其站供的用户中断供电。

5. 企业变电站

企业变电站是供大、中型企业专用的终端变电站，电压等级一般为 35～110 kV，进线为 1～2 回。

Program ❶ Overview of Electrical Equipment

Module 1 Overview of Power Plants and Substations

1.1.1 Power Plants

A power plant is an industrial plant that converts various natural energy sources (chemical, hydraulic, atomic, etc.) into electrical energy. According to the different energy sources used or the characteristics of converted energy, there are the following types of power plants.

1.1.1.1 Thermal power plants

A thermal power plant, or TPP, is a plant that converts the chemical energy of fossil fuels (coal, oil, natural gas, oil shale, etc.) into electrical energy. Most of the prime movers in thermal power plants are steam turbines, and gas turbines and diesel engines are also used. Thermal power plants can be further categorized as follows.

1. Condensing TPP

A schematic diagram of the production process of a condensing TPP is shown in Fig. 1-1. Pulverized coal is burned in the boiler furnace to heat the water in the boiler into superheated steam, which is sent to the steam turbine through the tubes, drives the steam turbine to rotate and change thermal energy into mechanical energy. The steam turbine drives the generator to rotate, and then changes the mechanical energy into electrical energy. The steam which has acted in the steam turbine is discharged into a condenser, into which circulating water is pumped by a circulating water pump, which condenses the discharged steam by cooling it rapidly. The condensate is sent by a condensate pump to a deaerator for deoxygenation (removal of gases, especially oxygen, from the water) and is then returned to the boiler by a feedwater pump.

The efficiency of a condensing TPP is low at 30%—40% because a large amount of heat is carried away by the circulating water in the condenser.

2. Cogeneration power plant

The production process of a cogeneration power plant is shown in Fig. 1-2. The difference between a cogeneration power plant and a condensing TPP is that a portion of the steam in the steam turbine that has acted is withdrawn from the middle section and supplied directly to the heat users, or, the water is heated by a heat exchanger and the hot water is supplied to the users. In this way, the heat carried away by the circulating water is reduced and efficiency is improved. Modern cogeneration power plants have efficiencies of 60%—70%.

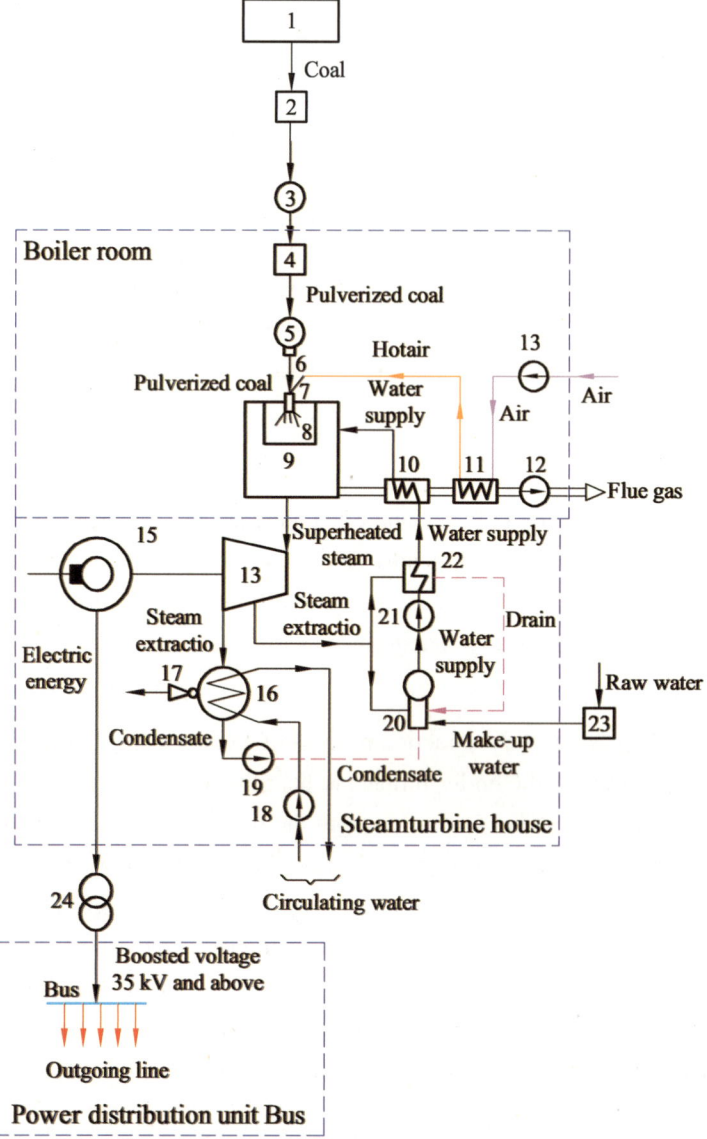

1—Coal yard; 2—Coal crusher; 3—Raw coal bunker; 4—Coal mill; 5—Pulverized coal bunker; 6—Pulverized coal feeder; 7—Burner; 8—Furnace; 9—Boiler; 10—Coal economizer; 11—Air preheater; 12—Induced draft fan; 13—Blowers; 14—Steam turbine; 15—Generator; 16—Condenser; 17—Extractor; 18—Circulating water pump; 19—Condensate pump; 20—Deaerator; 21—Feedwater pump; 22—Heater; 23—Water treatment equipment; 24—Step-up transformer.

Fig. 1-1 Schematic Diagram of the Production Process of a Condensing TPP

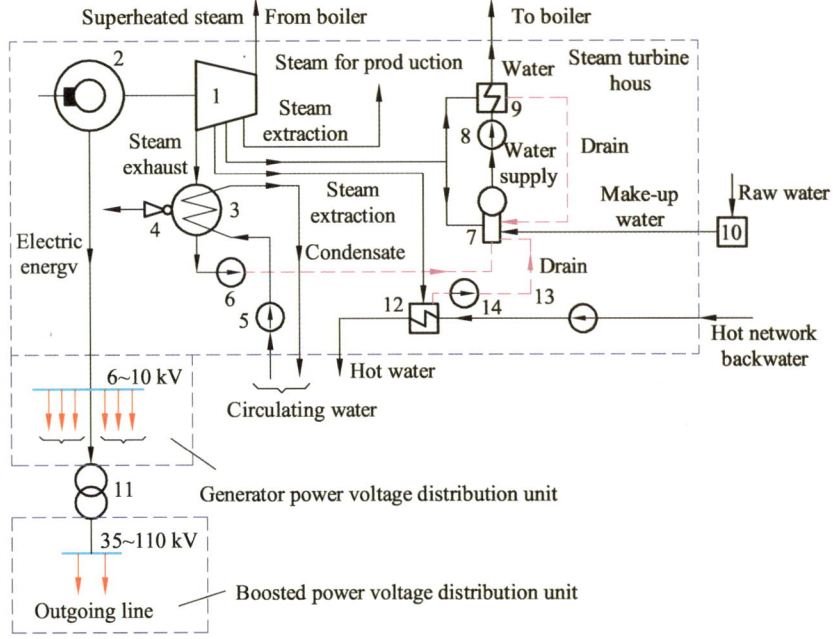

1—Steam turbine; 2—Generator; 3—Condenser; 4—Extractor; 5—Circulating water pump; 6—Condensate pump; 7—Deaerator; 8—Feedwater pump; 9—Heater; 10—Water treatment equipment; 11—Step-up transformer; 12—Heater; 13—Backwater pump; 14—Pump.

Fig. 1-2　Schematic Diagram of the Production Process of a Cogeneration Power Plant

Since the heat supply network cannot be too long, cogeneration power plants are always built in the vicinity of heat users. In addition, in order to maintain the high efficiency of the cogeneration power plant, it generally adopts the operation mode of "heat supply determining power generation", that is, when the heat load increases, the thermoelectric unit generates more electricity accordingly; when the heat load decreases, the thermoelectric unit generates less electricity accordingly. Therefore, its operation is not as flexible as that of condensing TPPs.

3. Gas turbine power plants

A power plant, that uses the gas turbine or the gas turbine and steam turbine in combined gas-steam cycle to drive a generator, is called gas turbine power plant. The former is generally used as a peak load modulation unit in power systems, while the latter is used to carry intermediate and base loads. This type of power plant can use either liquid or gaseous fuels. Gas turbines and combined cycle power generation fueled by natural gas have the advantages of high efficiency, low pollutant emissions, low initial investment, short construction period and easy load regulation, and have been rapidly developed in North America and Europe in recent years. At present, the stand-alone capacity of gas turbines has been developed to 300,000 kW.

The working principle of a gas turbine is similar to that of a steam turbine, except that instead of steam, the working medium is a high-temperature, high-pressure gas. The air is compressed and pressurized by the compressor and sent into the combustion chamber, and the fuel is pumped into the combustion chamber by the fuel pump. The high-temperature and high-pressure gas generated by combustion enters the gas turbine and expands to act, which pushes the gas turbine to rotate and drive power generation. The exhaust gas that has acted is discharged through the smokestack, or shunted partly for heating and cooling. This kind of power plant which only uses gas turbine to drive the generator has a thermal efficiency of only 35%—40%.

In order to improve thermal efficiency, a combined gas-steam cycle system is adopted, one of the models of which is shown in Fig. 1-3. The exhaust gas from the gas turbine enters the waste heat boiler, heats up the feedwater therein and produces high-temperature and high-pressure steam, which is sent to the steam turbine to act and drive the generator to generate electricity again; low-pressure steam is extracted from the steam turbine (when the generator stops generating electricity, the standby gas boiler is started to provide the source of steam), and hot and cold water is produced through the steam-type bromide refrigerator (lithium bromide is used as the absorbent) or the steam-water heat exchanger. This is a combined mode of electricity, heat and cooling. The combined cycle system have a thermal efficiency of 56%—85%.

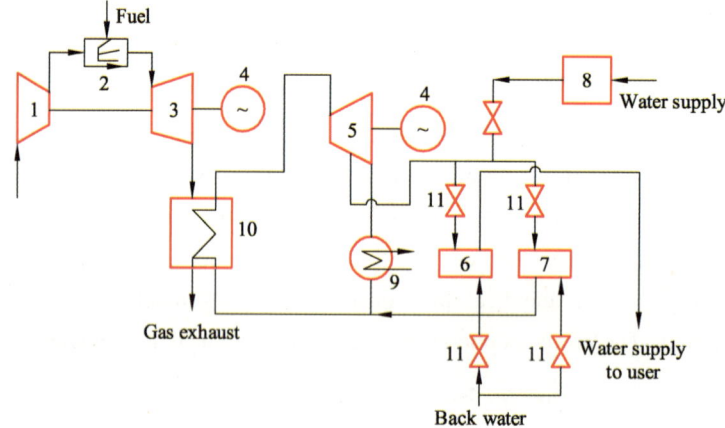

1—Compressor; 2—Combustion chamber; 3—Gas turbine; 4—Generator; 5—Steam turbine;
6—Steam-type bromide refrigerator; 7—Steam-water heat exchanger; 8—Standby gas boiler;
9—Condenser; 10—Waste heat boiler; 11—Cooling and heating changeover valve.

Fig. 1-3 Combined Gas-Steam Cycle System

1.1.1.2 Hydropower plants

Hydropower plant, or hydropower station, is a plant that converts potential energy and kinetic energy of water into electric energy. The prime mover of a hydropower station is water turbine. The water energy is converted into mechanical energy through water turbine, and then the water turbine drives the generator to convert the mechanical energy into electric energy.

1. Dam-type hydropower station

A barrage is built at the appropriate place of the river to form water reservoir, thereby raising the upstream water level and forming a large water level difference between upstream and downstream of the dam. This kind of hydropower station is called dam-type hydropower station. Dam-type hydropower station is suitable to be built in the river section with slow slope and large flow. According to the relative position of the powerhouse and dam, this kind of hydropower station can be divided into the several types below.

(1) Powerhouse at dam toe. A hydropower station at dam toe is shown in Fig. 1-4. The powerhouse is built behind the non-overflow dam section (downstream side) of the barrage, not bearing the pressure of water upstream, with the penstock passing through the dam body, and is suitable for high and medium heads.

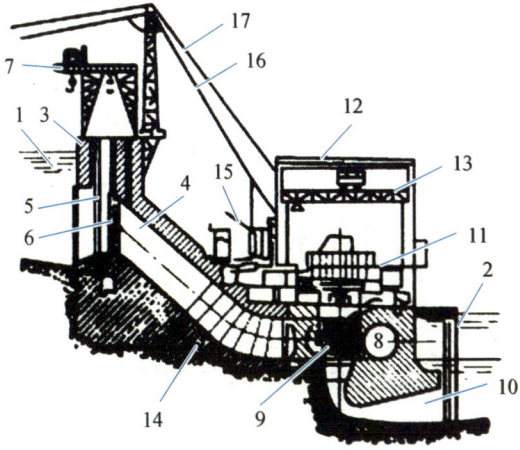

1—Upstream water level; 2—Downstream water level; 3—Dam; 4—Intake penstock; 5—Bulkhead gate; 6—Gate;
7—Crane; 8—Turbine spiral casing; 9—Turbine rotor; 10—Draft tube; 11—Generator; 12—Generator room; 13—Crane;
14—Generator voltage power distribution unit; 15—Step-up transformer;
16—Overhead line; 17—Lightning wire.

Fig. 1-4 Cross-section of a Hydropower Station at Dam Toe

The production process of a hydropower station is relatively simple: Generator is coaxially connected with turbine rotor, water enters into turbine spiral casing from upstream along penstock, impelling turbine rotor, and the turbine drives the generator to rotate, i.e., to send out electric energy; the water that has acted flows downstream through the draft tube; the electricity produced is stepped up by transformer and travels along the overhead line to the power distribution units outside the powerhouse, which in turn feeds into the power system.

(2) Overflow type powerhouse. An overflow type powerhouse is built behind the overflow dam section (downstream side), and the flood discharge flow is released from the top of the powerhouse to the downstream river channel. It is suitable for the situation with narrow river valley, large flood discharge flow from reservoir, and certain difficulty in layout of overflow and power generation zones.

(3) River-side powerhouse. A river-side powerhouse is built on the ground of the river bank downstream of the barrage, and the head race and penstock are laid on the ground or embedded underground.

(4) Underground powerhouse. An underground powerhouse has both the head race and the powerhouse built underground on the side of the dam.

(5) Powerhouse in dam. The penstock and powerhouse of the powerhouse in dam are built in the cavity of the concrete dam and are permanently located in the overflow dam section, which is suitable for narrow river valleys with high downstream flood water flow.

(6) Powerhouse in river channel. Hydropower stations with powerhouse in river channel are shown in Fig. 1-5. The powerhouse is connected with the barrage and becomes a part of the dam. The powerhouse bears the pressure of water and is suitable for the hydropower station with water head less than 50 m. The overflow dam and spillway in the figure are release structures for discharging flood and ensuring the safety of the dam.

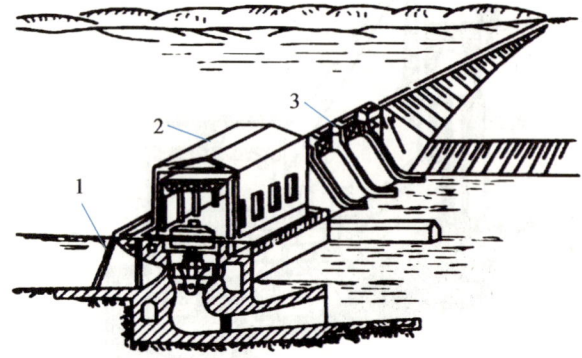

1—Inlet; 2—Powerhouse; 3—Waternotchdam.
Fig. 1-5 Schematic Diagram of Hydropower Station in River Channel

2. Diversion-type hydropower station

A hydropower station, which uses the diversion system to centralize the drop of natural river for power generation, is called a diversion-type hydropower station. Diversion-type hydropower station is suitable to be built in the river section with more curves or steeper river slope. With the shorter diversion system, the greater water head can be concentrated; it is also suitable for high-head hydropower station, thereby avoiding the construction of excessively high water retaining structures.

The diversion-type hydropower station is shown in Fig. 1-6. A low weir (low dam for water retention) is built at an appropriate section of the river, and water is introduced into the powerhouse through the diversion canal and the penstock, thus obtaining a large difference in water level.

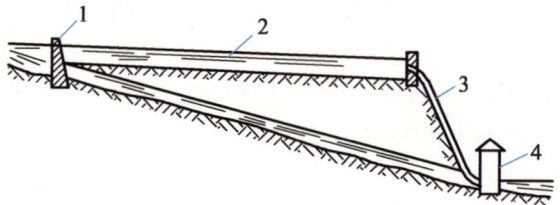

1—Low weir; 2—Diversion canal; 3—Penstock; 4—Powerhouse.
Fig. 1-6 Diversion-type Hydropower Station

3. Pumped storage power station

The hydropower station, which pumps water to high place for storage by utilizing the surplus power of the power system under the off-peak load and then releases the water to generate electricity under the peak load, is called pumped storage power station. It is a peak load shifting power source for the power system. In a power system dominated by thermal and nuclear power, the construction of an appropriate proportion of pumped storage power stations can improve the economy and reliability of system operation. Pumped storage power station can be dam type or diversion type.

The pumped storage power station is shown in Fig. 1-7. When the power system is in the valley load, its units are operated in the motor-pump mode, absorbing the active power of the power system to pump the downstream water to the upstream reservoir for storage, and converting the electric energy into water energy. At this point, it is the user. When the power system is at peak load, its units are operated in the turbine-generator mode, so that the stored water is used for power generation to meet the peak shifting needs. At this point, it's the power station.

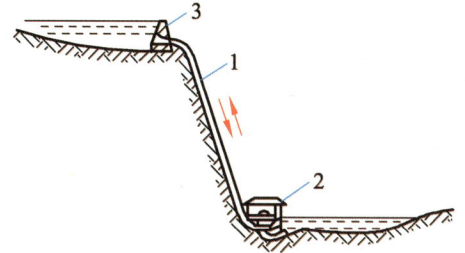

1—Water conveyance system; 2—Powerhouse and reservoir; 3—Upper reservoir.

Fig. 1-7　Pumped Storage Power Station

1.1.1.3　Nuclear power plants

A nuclear power plant is a power plant that converts the fission energy of atomic nuclei into electricity, and is fueled mainly by U235. U235 is prone to fission by the impact of slow neutrons, which releases a huge amount of energy, as well as releasing new neutrons. According to the moderator and coolant used, nuclear reactors can be categorized into light water reactors, heavy water reactors, graphite gas-cooled reactors and graphite boiling water reactors. Among them, light water reactors are divided into pressurized water reactors and boiling water reactors.

The production process of a nuclear power plant is similar to that of a general thermal power plant, i.e., the thermal energy generated by nuclear energy is utilized, and then the thermal energy is converted into mechanical energy and then into electrical energy in the same way as the power generation of a TPP. Light water reactor (pressurized water reactor and boiling water reactor) nuclear power plants are the most numerous among nuclear power plants. A schematic diagram of the power generation method of light water reactor type nuclear power plants is shown in Fig. 1-8.

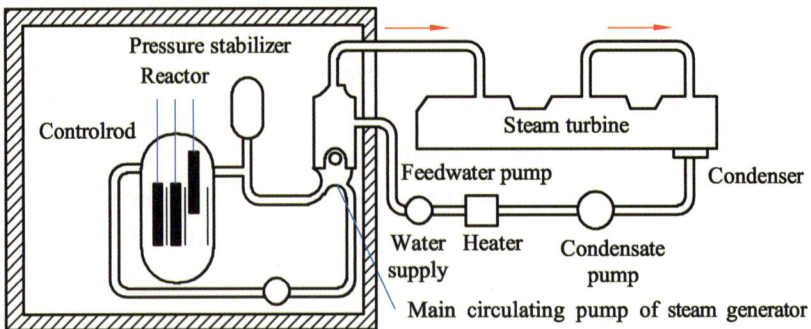

(a) Pressurized water reactor type nuclear power plant

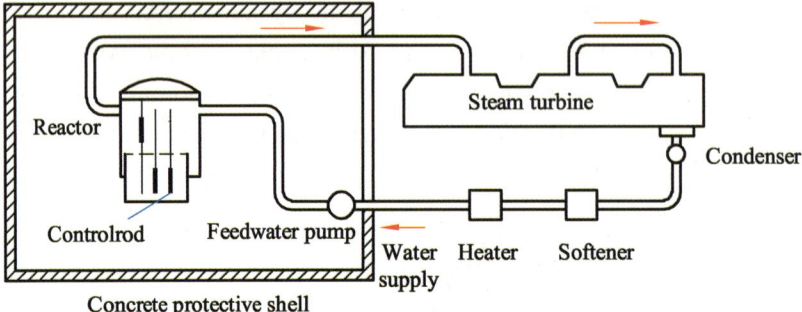

(b) Boiling water reactor type nuclear power plant

Fig. 1-8　Schematic Diagram of Power Generation Method of Light Water Reactor Type Nuclear Power Plants

Pressurized water reactor type nuclear power plants actually replace the boilers of a typical thermal power plant with a nuclear reactor and steam generator. There are typically more than 100 to 200 fuel assemblies in the reactor. Under the action of the main circulating water pump (also known as pressurized water reactor coolant pump or main pump), distilled water with a pressure of 15.2—15.5 MPa and a temperature of about 290 °C continuously circulates in the left circuit (i.e., the primary circuit, with 2—4 parallel circuits), is heated up to about 320 °C when passing through the reactor, then enters the steam generator, and transmits its own heat to the feedwater in the right circuit (i.e., the secondary circuit), turning it into saturated or slightly superheated steam. The steam enters the steam turbine along the pipe and expands to act, pushing the steam turbine and driving the generator to produce electricity. The working process of the secondary circuit is similar to that of a thermal power plant.

The control of rapid-change reactivity in pressurized water reactors is achieved mainly by changing the position of the control rods (neutron absorbers with silver-indium-cadmium material inside) in the core.

The role of the pressure stabilizer (with safety and relief valves) in the left circuit is to be used for system step-up (force) during plant startup and to automatically regulate system pressure and water level during normal operation and to provide overpressure protection.

Boiling water reactor type nuclear power plants use boiling light water as the slow water agent and coolant and generate saturated steam directly in the reactor, which is fed into the steam turbine to act and generate electricity. The exhaust steam from the steam turbine is condensed, purified by a lightener, heated by a heater, and then fed to the reactor by a feedwater pump.

1.1.1.4　New energy generation

1. Wind power generation

The energy possessed by flowing air is called wind energy. Globally, the available wind energy is about $2\times10^6\,\mathrm{kW}$. Wind energy is a renewable energy source, a process energy source that cannot be directly stored and is random, which adds technical complexity to the utilization of wind energy.

The method of power generation, that converts wind energy to electric energy, is called wind power generation. In area rich in wind energy, wind turbine generator system is installed in groups in a certain arrangement to form a cluster, which is called wind power plant. Their units can be as many as dozens, hundreds, or even thousands of units, which is an effective form of large-scale development and utilization of wind energy.

The wind power generation unit is shown in Fig. 1-9. Wind turbine converts wind energy into mechanical energy (belonging to low-speed rotating machinery), and the speed increaser gearbox changes the low-speed rotation on the wind turbine shaft into high-speed rotation, driving the generator to send out electrical energy; the electrical energy is led through the cable line to the power distribution unit, and then fed into the power grid.

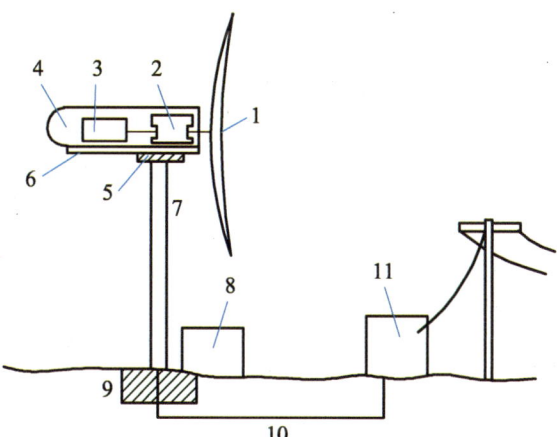

1—Wind turbine; 2—Speed increaser gearbox; 3—Generator; 4—Control system; 5—Drive unit; 6—Bottom slab and cover; 7—Tower; 8—Control and protection devices; 9—Civil foundations; 10—Cable routes; 11—Power distribution unit.

Fig. 1-9　Wind Turbine

Most of the blades (2—3 blades) of wind turbines are made of polyester resin FRP material. The tower is made of steel (conical barrel type or truss type). The speed increaser gearbox is generally a three-stage gear drive. The stand-alone capacity of wind turbines ranges from a few tens of watts to several megawatts, and wind turbines above 100 kW are synchronous or asynchronous generators. Large- and medium-sized wind turbines are equipped with control systems composed of microcomputers or PLCs to realize control, self-inspection, display and other functions.

2. Ocean energy generation

Ocean energy is the renewable energy contained in seawater, such as tidal energy, wave energy, ocean current energy, ocean thermal energy and ocean salinity energy. Tidal power generation refers to the power generation that uses the potential energy of the tide to generate electricity, namely that dam is constructed at the inlet of bay or the estuary with high tidal range to form a reservoir, water-turbine generator set is installed in the dam or at the dam side, and the tidal range in both sides of the dam is used in driving the water-turbine generator set to generate electricity.

(1) Single-reservoir unidirectional type. A single-reservoir unidirectional type tidal power station is shown in Fig. 1-10. This type of power station is built with only one reservoir and installed with a unidirectional turbine generator set (the generator is mounted in a sealed bulb body), which generates electricity at low tide. When the rising tide to the water level in the reservoir, it opens the gate to fill the reservoir; until the water level is flush at a high level inside and outside the reservoir, the gate closes and waits for the tide to gradually fall. When the difference between the water level inside and outside the reservoir reaches the head of the unit to start, it opens the gate to generate electricity (when the water level in the reservoir gradually falls); until the difference between the water level inside and outside the reservoir is less than the minimum head of the unit required for generating electricity, it again closes the gate and waits for it, and then transfers to the next cycle.

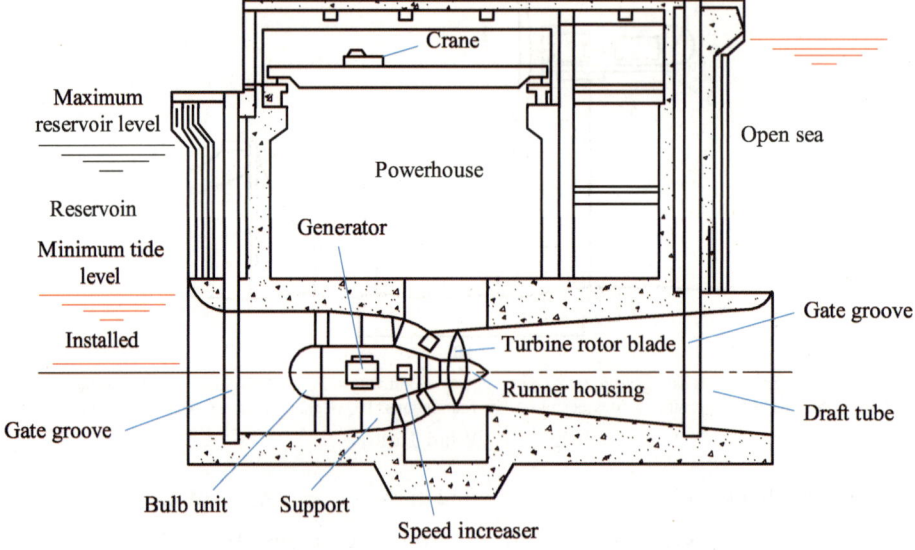

Fig. 1-10 Single-Reservoir Unidirectional Type Tidal Power Station

(2) Single-reservoir bi-directional type. A single-reservoir bi-directional type tidal power station is shown in Fig. 1-11. This type of power station is also built with only one reservoir and installed with bi-directional turbine generator sets, and generates electricity at both high and low tides. When the tide rises to a certain height, gates A and B are opened to introduce tidal water into the station to impulse the units to generate electricity; when the tide about to end, gates E and F are quickly opened; when the reservoir is full, the gates are closed; when the tide falls to a certain water level difference, gates C and D are opened to impulse the unit again to generate electricity. The ebb-and-flow two-way power generation has been realized.

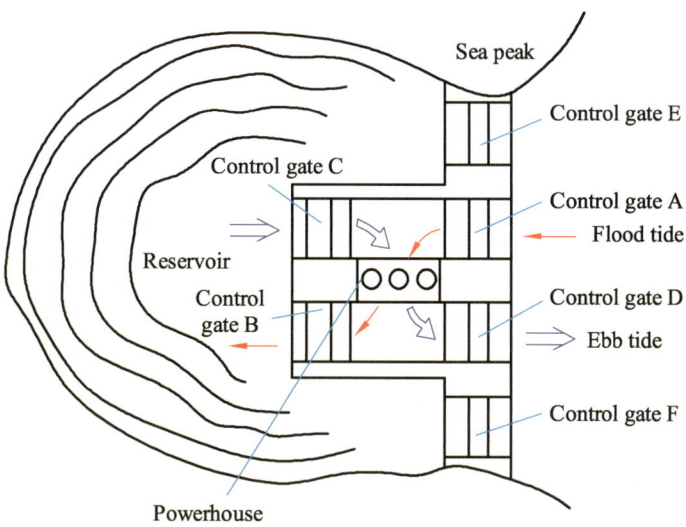

Fig. 1-11   Single-Reservoir Bi-directional Type Tidal Power Station

(3) Double-reservoir (high and low reservoirs) type. Two adjacent reservoirs are constructed, and water-turbine generator set is installed in the separation dam between the two reservoirs. The high reservoir is equipped with an intake gate, thus water can flow in the reservoir when the tide level is higher than the water level in the reservoir (there is no water flowing in the low reservoir), thereby maintaining the high water level to the greatest extent; the low reservoir is equipped with a water release gate, thereby ensuring water release when the tide level is lower than the water level in the reservoir. In this way, there is a water level difference between the two reservoirs all day long, thereby ensuring continuous power generation.

3. Geothermal power generation

A power generation, that uses the thermal energy resources inside the Earth such as underground steam or hot water, is called geothermal power generation. Currently, the maximum stand-alone capacity of geothermal power generation is 150,000 kW. The principle and equipment of geothermal steam power generation are basically the same as those of thermal power plant. There are two basic types of power generation with underground hot water.

(1) Flash evaporation geothermal power generation system (also known as the decompression and expansion method). This method is to make the underground hot water into low-pressure steam for the steam turbine to act, as shown in Fig. 1-12. The underground hot water is deaerated by a deaerator, and then enters the first-stage flash tank for decompression and expansion to produce primary steam (about 10% of the hot water), which is fed into the high-pressure part of the steam turbine to act. The remaining hot water enters the second-stage flash tank, and then carries out the second decompression and expansion to produce secondary steam. Since its pressure is lower than that of the first stage, it is sent to the low-pressure part of the steam turbine to act. The actual expansion stages generally do not exceed four stages. In China, the geothermal power station in Yangbajing is a two-stage expansion station.

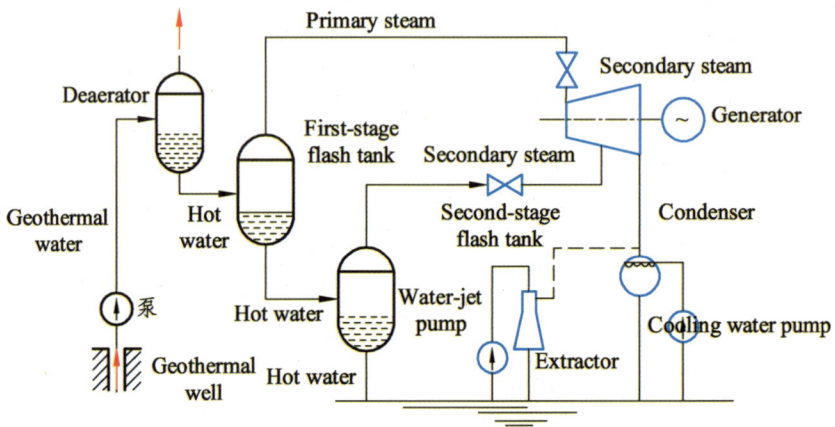

Fig. 1-12 Flash Evaporation Geothermal Power Generation System

Expansion evaporation is also known as flash evaporation. When geothermal water with a certain pressure and temperature is injected into a container with lower pressure, a part of the hot water is rapidly vaporized into steam due to the water temperature being higher than the saturation temperature of the container pressure, and the temperature is lowered until both the water and the steam are saturated at that pressure. When the fluid at the geothermal wellhead is wet steam, it first enters the vapor-water separator. The separated steam is sent to the steam turbine, and the remaining water then enters the flash tank.

(2) Double-cycle geothermal power generation system (also known as the intermediate medium method). The process is shown in Fig. 1-13. Underground hot water is pumped to the evaporator of the power station by a deep well pump to heat some kind of low-boiling point working medium (such as Freon, iso-butane, n-butane), so that it turns into a low-boiling point working medium steam, which pushes the turbine generator to generate electricity; the exhaust steam of the steam turbine is condensed into a liquid through the condenser, and is pumped back to the evaporator by the working medium pump to be reheated and circulated for use. In order to fully utilize the waste heat of the geothermal

water, the geothermal water discharged from the evaporator is preheated by a preheater, and then the low-boiling point working medium liquid from the condenser. The hot water and working medium of this system constitute independent systems respectively, so it is called double cycle system.

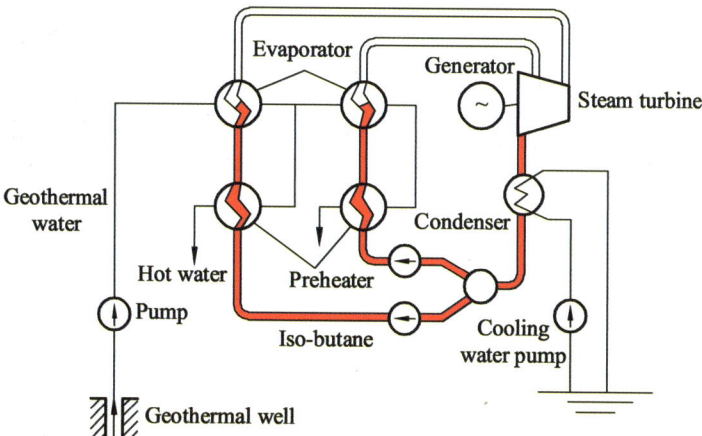

Fig. 1-13　Double-Cycle Geothermal Power Generation System

4. Solar power generation

Solar energy is the electromagnetic radiation energy emitted from the Sun into the cosmic space, and the solar energy that reaches the Earth's surface is $8.2\times10^9\,\text{kW}$, with an energy density of about 1kW/m^2. There are two forms of solar power generation, namely thermal power generation and photovoltaic power generation.

(1) Solar thermal power generation. Solar thermal power generation is the installation that converts the absorbed solar radiant heat energy into electric energy. Its basic composition is similar to that of regular thermal power equipment. It is also divided into centralized and distributed types.

Centralized solar thermal power generation is also known as tower solar thermal power generation, and its thermodynamic system flow is as shown in Fig. 1-14. It is a large array of heliostats (mirrors) neatly laid out over a large area of the site, and each is equipped with a tracking system. It accurately reflects and concentrates sunlight into a heat absorber (also called a receiver) on top of a tall tower, converts the absorbed light energy into heat energy, and turns the working medium (water) in the heat absorber into steam, which is sent to the steam turbine through pipelines to drive the unit to generate electricity.

Distributed solar thermal power generation is the installation of many sets of small solar collectors of the same structure on a large site, the heat energy generated by each set of devices is pooled together through pipelines, and thermoelectric conversion is carried out to generate electricity.

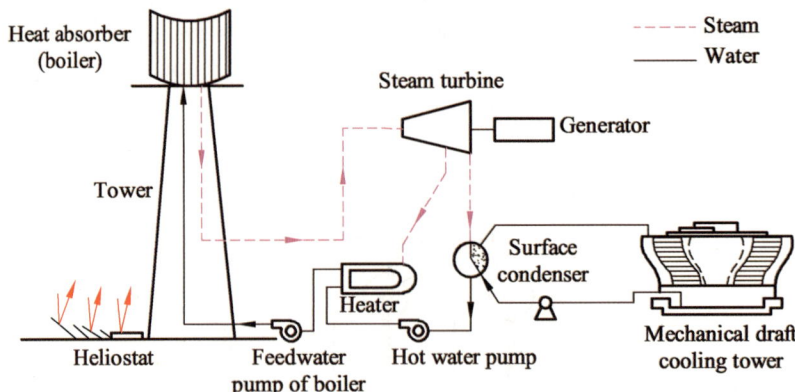

Fig. 1-14 Thermal System Flow of Tower Solar Power Station

(2) Solar photovoltaic power generation. Solar photovoltaic power generation is the direct conversion of the Sun's light energy into electric energy without a thermal process, and there are various ways of generating electricity, of which photovoltaic power generation is the mainstream. Photovoltaic power generation is the direct conversion of light irradiated to solar cells (also known as photovoltaic cells, a semiconductor device that produces a volta effect when irradiated by light) into electrical energy output.

5. Biomass power generation

Biomass energy is the energy stored in the biomass of green plants by converting solar energy to chemical energy through chlorophyll, which belongs to renewable energy. Fuel wood, crop straw, human and animal manure, organic waste, and industrial organic wastewater are the main biomass energy resources. The biomass power production system is a power generation project based on biomass energy, including waste incineration power generation, biogas power generation, and bagasse power generation.

6. Magneto-hydrodynamic power generation

Magneto-hydrodynamic power generation, also known as plasma power generation, is to make extremely high-temperature and highly ionized gas flow through a strong magnetic field at high speed (1,000 m/s) and generate electricity directly. At this time, the electrons in the gas collide with each other by the magnetic force and the activated metal particles (potassium and cesium) in the gas, and flow to the electrodes perpendicular to the magnetic lines of force and generate direct current.

1.1.2 Substations

A substation is an intermediate link between the power plant and the user, playing the role of transforming and distributing electric energy. There are various ways to categorize substations, which can be classified as per the voltage class, step-up or step-down voltage, and position in the power system. According to the position of the substation in the system, it can be divided into the following categories. Schematic diagrams of the various types of substations are shown in Fig. 1-15.

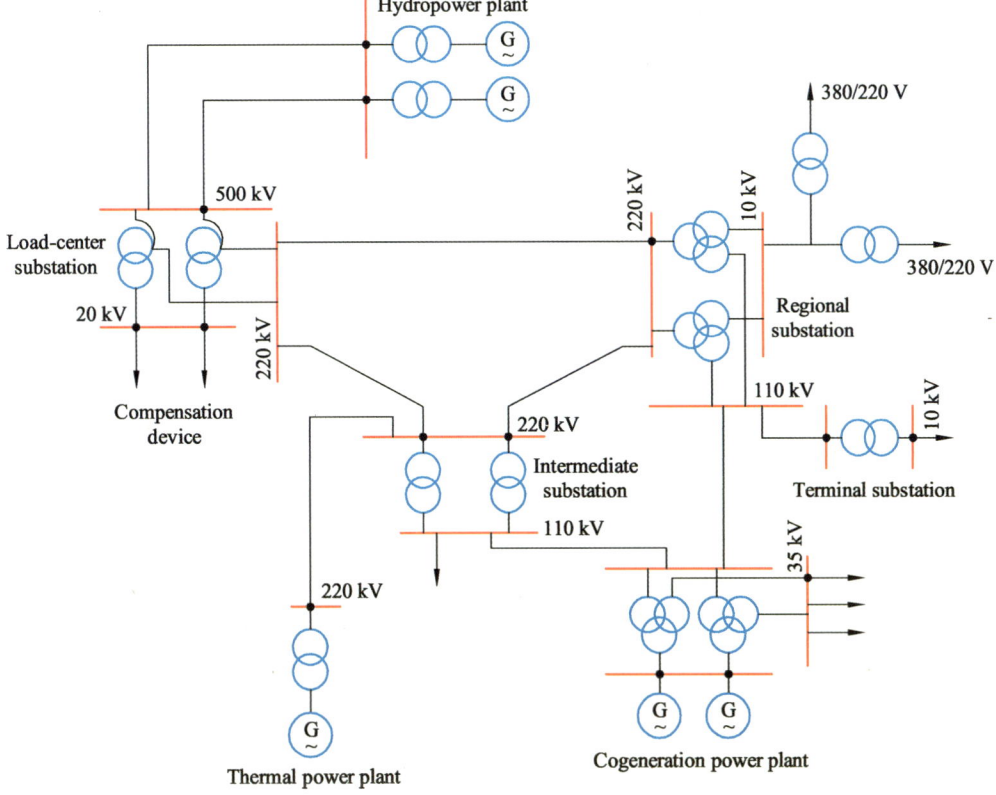

Fig. 1-15　Wiring Diagram of Power System

1. Load-center substation

A load-center substation is located at the center node of the power system, connecting several high- and medium-voltage parts of the power system, collecting multiple power supplies and multiple high-capacity liaison lines, with large substation capacity and voltage (referring to the HV side, the same as below) class of 330—500 kV. The power failure of the whole substation can cause a system stepout and even breakdown.

2. Intermediate substation

An intermediate substation is generally located in the main loop line of the system or at the interface of the main trunk line of the system, collecting 2 to 3 power supplies. It is dominated by current switching on the HV side, and the current is also stepped down to supply the local users. The substation mainly plays the role of the intermediate link, with a voltage class of 220—330 kV. The power failure of the whole substation can cause a regional grid stepout.

3. Regional substation

A regional substation mainly supplies power to regional users and is the main substation in a region or a city, with a general voltage class of 110—220 kV. The power failure of the whole substation can only cause an interruption of the power supply in this region.

4. Terminal substation

A terminal substation is located at the terminal of the transmission line and is close to the load point. After step-down, it can supply power to the user directly and does not undertake the power transfer task, with a voltage class of 110 kV and below. The power failure of the whole substation can only cause an interruption of power supply to its users.

5. Enterprise substation

An enterprise substation is a special terminal substation for large- and medium-sized enterprises, with a general voltage class of 35—110 kV and an incoming line with 1—2 circuits.

模块二　主要电气设备

为了满足电能生产、输送和分配的需要，发电厂和变电站安装有各种电气设备，用于实现起动、转换、监视、测量、调整、保护、切换和停止等操作。按电压等级可分为高压电器和低压电器；按所起的作用不同电气设备可分为一次设备和二次设备两大类。

一、一次设备

直接生产、转换和输配电能的设备，称为一次设备，主要有以下几种。

1. 生产和转换电能的设备

生产和转换电能的设备有同步发电机、变压器及电动机，它们都是按电磁感应原理工作的，统称为电机。

2. 开关电器

开关电器的作用是接通或断开电路。高压开关电器主要有以下几种。

（1）断路器（俗称开关）。断路器有灭弧装置，可用来接通或断开电路的正常工作电流、过负荷电流或短路电流，是电力系统中最重要的控制和保护电器。

（2）隔离开关（俗称刀闸）。隔离开关没有灭弧装置，用来在检修设备时隔离电源，进行电路的切换操作及接通或断开小电流电路。它一般只有在电路断开的情况下才能操作。在各种电气设备中，隔离开关的使用量是最多的。

（3）负荷开关。负荷开关具有简易的灭弧装置，可以用来接通或断开电路的正常工作电流和过负荷电流，还可用来在检修设备时隔离电源，但不能用来接通或断开短路电流。

另外，还有用于配电系统的自动重合器和自动分段器等。

低压开关电器包括刀开关、组合开关和低压断路器等。

3. 限流电器

限流电器包括串联在电路中的普通电抗器和分裂电抗器，其作用是限制短路电流，使发电厂或变电站能选择轻型电器和选用截面积较小的导体。

4. 载流导体

载流导体包括母线、架空线和电缆线等。母线用来汇集和分配电能或将发电机、变压器与配电装置连接；架空线和电缆线用来传输电能。

5. 补偿设备

（1）调相机。调相机是一种不带机械负荷运行的同步电动机，主要用来向系统输出感性无功功率，以调节电压控制点或地区的电压。

（2）电力电容器。电力电容器补偿有并联和串联补偿两类。并联补偿是将电容器与用电设备并联，它发出无功功率，供给本地区需要，避免长距离输送无功，减少线路电能损耗和电压损耗，提高系统供电能力；串联补偿是将电容器与线路串联，抵消系统的部分感抗，提高系统的电压水平，也相应地减少系统的功率损失。

（3）消弧线圈。消弧线圈用来补偿小接地电流系统的单相接地电容电流，以利于熄灭电弧。

（4）并联电抗。并联电抗器一般装设在 330 kV 及以上超高压配电装置的某些线路侧。其作用主要是吸收过剩的无功功率，改善沿线电压分布和无功分布，降低有功损耗，提高送电效率。

6. 互感器

互感器包括电流互感器和电压互感器。电流互感器作用是将交流大电流变成小电流（5 A 或 1 A），供电给测量仪表和继电保护装置的电流线圈。电压互感器作用是将交流高电压变成低电压（100 V 或 $100/\sqrt{3}$ V），供电给测量仪表和继电保护装置使用。它们使测量仪表和保护装置标准化和小型化，使测量仪表和保护装置等二次设备与高压部分隔离，且互感器二次侧均接地，从而保证设备和人身安全。

7. 保护电器

保护电器包括用于过负荷电流或短路电流保护的熔断器（俗称保险）和防御过电压的设备，即防雷装置。

熔断器用来断开电路的过负荷电流或短路电流，保护电气设备免受过载和短路电流的危害。熔断器不能用来接通或断开正常工作电流，必须与其他电器配合使用。

防雷装置包括：避雷器、避雷针、避雷线（架空地线）、避雷带和避雷网等。

8. 绝缘子

绝缘子用来支持和固定载流导体，使载流导体与地绝缘，或使装置中不同电位的载流导体间绝缘。

9. 接地装置

接地装置用来保证电力系统正常工作或保护人身安全,前者称工作接地,后者称保护接地。

二、二次设备

对一次设备进行监察、测量、控制、保护、调节的辅助设备,称为二次设备。

1. 测量表计

测量表计用来监视、测量电路的电流、电压、功率、电能、频率及设备的温度等,如电流表、电压表、功率表、电能表、频率表、温度表等。

2. 绝缘监察装置

绝缘监察装置用来监察交、直流电网的绝缘状况。

3. 控制和信号装置

控制主要是指采用手动(用控制开关或按钮)或自动(继电保护或自动装置)方式通过操作回路实现配电装置中断路器的分、合闸。断路器都有位置信号灯,有些隔离开关有位置指示器。主控制室设有中央信号装置,用来反映电气设备的事故或异常状态。

4. 继电保护及自动装置

继电保护的作用是当发生故障时,作用于断路器跳闸,自动切除故障元件;当出现异常情况时发出信号。自动装置的作用是实现发电厂的自动并列、发电机自动调节励磁、电力系统频率自动调节、按频率启动水轮机组;实现发电厂或变电站的备用电源自动投入、输电线路自动重合闸及按事故频率自动减负荷等。

5. 直流电源设备

直流电源设备包括蓄电池组和硅整流装置,用作开关电器的操作、信号、继电保护及自动装置的直流电源,以及事故照明和直流电动机的备用电源。

6. 塞流线圈(又称高频阻波器)

塞流线圈是电力载波通信设备中必不可少的组成部分,它与耦合电容器、结合滤波器、高频电缆、高频通信机等组成电力线路高频通信通道。塞流线圈起到阻止高频电流向变电站或支线泄漏、减小高频能量损耗的作用。

Module 2　Main Electrical Equipment

In order to meet the needs of the production, transmission, and distribution of electric energy, power plants and substations are installed with a variety of electrical equipment, to realize the startup, conversion, monitoring, measurement, adjustment, protection, switching, and stopping and other operations. According to the voltage class, it can be divided into high-voltage and low-voltage electrical equipment; according to the different roles, electrical equipment can be divided into two categories: primary equipment and secondary equipment.

1.2.1　Primary Equipment

Equipment that directly produces, converts, transmits, and distributes electrical energy is known as primary equipment and includes the following.

1. Equipment for production and conversion of electrical energy

Equipment for production and conversion of electrical energy includes synchronous generators, transformers, and electric motors. They all work on the principle of electromagnetic induction and are collectively known as motors.

2. Switching devices

The function of switching devices is to connect or disconnect circuits. There are mainly the following types of HV switching devices.

(1) Circuit breaker (commonly known as a switch). A circuit breaker has an arc extinguishing device, that can be used to connect or disconnect the normal operating current, overload current, or short-circuit current of the circuit, which is the most important control and protection device of the power system.

(2) Disconnector (commonly known as a knife switch). A disconnector does not have an arc extinguishing device and is used to isolate the power supply during maintenance, to switch circuits, and to connect or disconnect small current circuits. It can generally only be operated when the circuit is open. Disconnectors are used most frequently in a variety of electrical equipment.

(3) Load switch. Load switches have a simple arc extinguishing device that can be used to connect or disconnect the normal operating current and overload current of a circuit, but cannot be used to connect or disconnect short-circuit currents. It can also be used to isolate the power supply during maintenance.

There are also automatic reclosers and automatic sectionalizers for power distribution systems. LV switching devices include knife switches, combination switches, and LV circuit breakers.

3. Current-limiting devices

Current-limiting devices include ordinary reactors and disruptive reactors connected in series in the circuit, whose function is to limit the short-circuit current, so that the power plant or substation can choose light devices and conductors with a smaller cross-sectional area.

4. Current carrying conductor

It includes a bus, overhead line, and cable line. A bus is used to collect and distribute power or connect generators, transformers, and distribution units; overhead lines and cable lines are used to transmit power.

5. Compensation equipment

(1) Phase modifier. A phase modifier is a synchronous motor that operates without mechanical load and is mainly used to output inductive reactive power to the system to regulate the voltage at the voltage control point or area.

(2) Power capacitor. The compensation of power capacitors is divided into two categories: shunt and cascade compensation. Shunt compensation involves connecting the capacitor in parallel with the power consumer equipment. It sends out reactive power to supply the needs of the region, avoids long-distance transmission of reactive power, reduces line power loss and voltage loss, and improves the power supply capacity of the system. Cascade compensation is to connect the capacitor in series with the line to offset part of the inductive reactance of the system, improve the voltage level of the system, and also reduce the power loss of the system accordingly.

(3) Arc suppression coil. An arc suppression coil is used to compensate for the single-phase grounding capacitive current of the small grounding current system, to facilitate suppressing the arc.

(4) Shunt reactor. Shunt reactors are generally installed in the 330 kV and above UHV power distribution unit of some line sides. Its role is mainly to absorb excess reactive power, improve the voltage distribution and reactive power distribution along the line, reduce active losses, and improve the efficiency of power transmission.

6. Transformer

It includes current and voltage transformers. The role of the current transformer is to turn high AC current into a small current (5 A or 1 A), and supply power to the current coils of the measuring instruments and relay protection devices; the role of the voltage transformer is to turn high AC voltages into low voltages (100 V or $100/\sqrt{3}$ V) to supply the voltage coils of the measuring instruments and relay protection devices. They standardize and miniaturize measuring instruments and protection devices, isolate secondary equipment such as measuring instruments and protection devices from high-voltage parts, and make the secondary side of the transformer grounded, thus ensuring the safety of equipment and people.

7. Protective devices

Protective devices include fuses for overload current or short-circuit current protection and equipment for defense against overvoltage, i.e., lightning protection devices.

Fuses are used to disconnect the overload current or short-circuit current of a circuit and to protect electrical equipment from overload and short-circuit currents. Fuses cannot be used to connect or disconnect the normal operating current, and must be used in conjunction with other electrical devices.

Lightning protection devices include lightning arresters, lightning rods, lightning wires (overhead ground wires), lightning strips, and lightning conduction.

8. Insulator

Insulators are used to support and secure current carrying conductors and to insulate current carrying conductors from the ground or between current carrying conductors of different potentials in an installation.

9. Grounding device

The grounding device is used to ensure the normal operation of the power system or to protect personal safety. The former is called working grounding, and the latter is called protective grounding.

1.2.2 Secondary Equipment

Auxiliary equipment for monitoring, measuring, controlling, protecting, and regulating primary equipment is called secondary equipment.

1. Measuring meters

Measuring meters are used to monitor and measure the current, voltage, power, electrical energy, frequency, and temperature of the circuit, including an ammeter, voltmeter, power meter, electrical energy meter, frequency meter, and thermometer.

2. Insulation monitoring devices

Insulation monitoring devices are used to monitor the insulation condition of AC and DC networks.

3. Control and signaling devices

Control refers primarily to the opening and closing of circuit breakers in a power distribution unit by means of operating circuits, either manually (with control switches or pushbuttons) or automatically (with relays or automatic devices). Circuit breakers are provided with position signaling lamps and some disconnectors have position indicators. The main control room is equipped with a central signaling device to reflect the accidental or abnormal status of electrical equipment.

4. Relay protection and automatic devices

The function of relay protection is to act on the tripping of the circuit breaker when a fault occurs, automatically remove the faulty components, and send out a signal when an abnormal situation occurs. The function of the automatic device is used to realize the automatic parallelism, the automatic excitation adjustment of the generator, the automatic adjustment of the frequency of the power system, and the starting of the hydraulic turbine set according to the frequency of the power plant; and to realize the automatic input of the standby power supply, the automatic reclosing of the transmission line, and the automatic load shedding according to the frequency of the accident of the power plant or the substation.

5. DC power equipment

DC power equipment includes battery packs and silicon rectifiers used as DC power sources for switching devices for operation, signaling, relay protection, and automation, as well as backup power sources for emergency lighting and DC motors.

6. Plug-flow coil (also known as a high-frequency wave trapper)

Plug-flow coil is an indispensable component of the power line carrier communication equipment. It forms a high-frequency communication channel of power lines with coupling capacitors, combined filters, high-frequency cables, and high-frequency communication equipment. The plug-flow coil plays a role in preventing the leakage of high-frequency current to the substation or branch line and reducing high-frequency energy loss.

模块三　电气设备符号及装置概述

一、电气设备的符号

图形符号是用于表示电气图中电气设备、装置、元器件的一种图形和符号。文字符号是电气图中电气设备、装置、元器件的种类字母和功能字母代码。文字符号的字母应采用大写的拉丁字母。文字符号分为基本文字符号和辅助文字符号两种。

二、电气主接线和配电装置的概念

1. 电气主接线

一次设备按预期的生产流程所连成的电路，称为电气主接线。主接线表明电能的生产、汇集、转换、分配关系和运行方式，是运行操作、切换电路的依据，又称一次接线、一次电路主系统或主电路。用国家规定的图形和文字符号表示主接线中的各元件，并依次连接起来的单线图，称电气主接线图。某火电厂的电气主接线图如图 1-16 所示。

该电厂有两个电压等级，即发电机电压 10 kV 及升高电压 110 kV；W1～W3 是发电机电压母线。工作母线由断路器 QFd（称分段断路器）分为 W1 和 W2 两段，备用母线 W3 不分段；W4、W5 为升高电压母线；断路器 QFc 起到联络两组母线的作用，称母线联络断路器（简称母联断路器）；每回进出线都装有断路器和隔离开关，断路器母线侧的隔离开关称母线隔离开关，断路器线路侧的隔离开关称线路隔离开关；发电机 G1 和 G2 发出的电力送至 10 kV 母线，一部分电能由电缆线路供给近区负荷，剩余电能则通过升压变压器 T1 和 T2 送到升高电压母线 W4、W5；各电缆馈线上均装有电抗器 L，以限制短路电流；由于 G1 和 G2 足够供给本地区负荷，所以发电机 G3 不再接在 10 kV 母线上，而与变压器 T3 单独接成发电机-变压器单元，以减少发电机电压母线及馈线的短路电流。

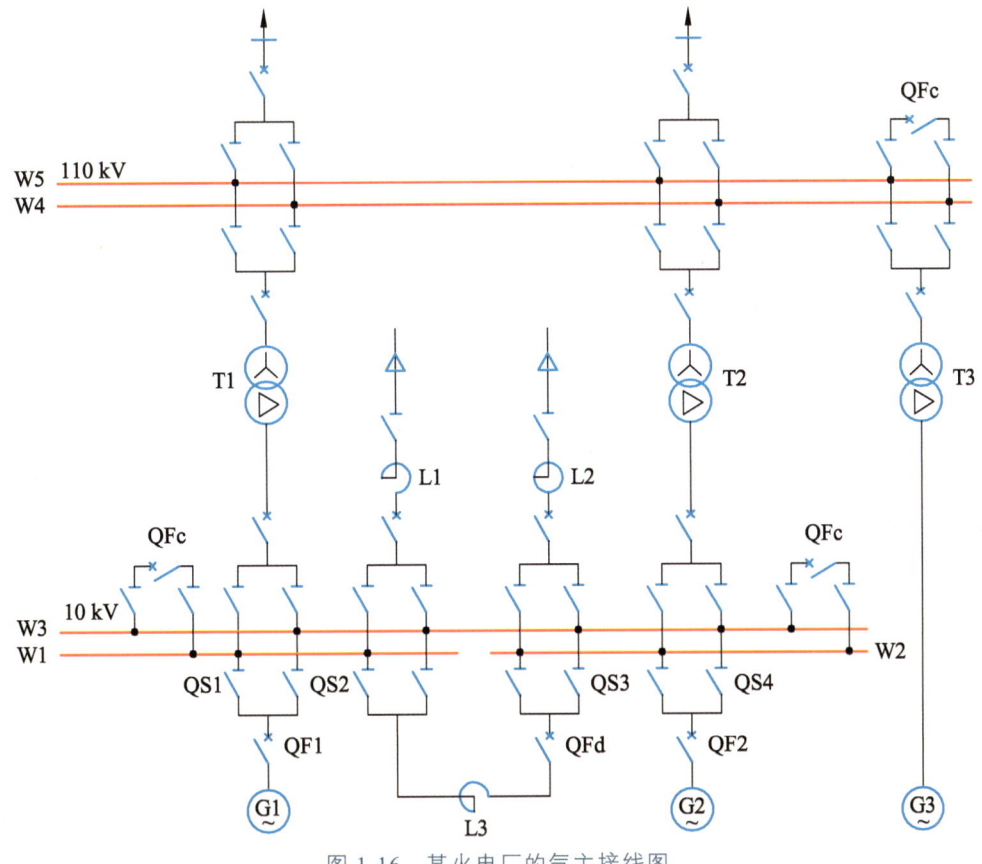

图 1-16　某火电厂的气主接线图

发电厂和变电站的主接线，是根据容量、电压等级、负荷等情况设计，并经过技术经济比较后选出最佳方案。

2. 配电装置

按主接线图，由母线、开关设备、保护电器、测量电器及必要的辅助设备组建成接受和分配电能的电工建筑物，称为配电装置。配电装置是发电厂和变电站的重要组成部分。

配电装置按电气设备的安装地点可分为以下两种：

（1）屋内配电装置。全部设备都安装在屋内。

（2）屋外配电装置。全部设备都安装在屋外（即露天场地）。

按电气设备的组装方式可分为以下两种：

（1）装配式配电装置。电气设备在现场（屋内或屋外）组装。

（2）成套式配电装置。制造厂预先将各单元电路的电气设备装配在封闭或不封闭的金属柜中，构成单元电路的分间。成套配电装置大部分为屋内型，也有屋外型。

配电装置还可按其他方式分类，例如按电压等级分类，称 10 kV 配电装置、35 kV 配电装置、110 kV 配电装置、220 kV 配电装置、500 kV 配电装置等。

Module 3 Overview of Electrical Equipment Symbols and Devices

1.3.1 Symbols of Electrical Equipment

Graphic symbols are a kind of graphics and symbols used to represent electrical equipment, devices, and components in electrical diagrams. Text symbols are alphabetic and functional alphabetic codes for types of electrical equipment, devices, and components in electrical diagrams. The letters of the text symbols should be capitalized in latin letters. Text symbols are divided into basic text symbols and auxiliary text symbols.

1.3.2 Concepts of Main Electrical Wiring and Power Distribution Units

1. Main electrical wiring

The circuit formed by the primary equipment in accordance with the expected production process is called the main electrical wiring. The main wiring indicates the production, convergence, conversion, distribution relations, and mode of operation of electrical energy, which is the basis for running operations and switching circuits. It also known as the primary connection, the primary circuit of the main system, or the main circuit. A single-line diagram that represents the components in the main electrical wiring and connects them in sequence with the nationally specified graphic and text symbols is called a main electrical wiring diagram. The main electrical wiring diagram of a thermal power plant is shown in Fig. 1-16.

The plant has two voltage classes, i.e. generator voltage 10 kV and boosted voltage 110 kV; W1-W3 are generator voltage buses. The working bus is divided into two sections W1 and W2 by circuit breaker QFd (called section circuit breaker), and the standby bus W3 is not segmented; W4 and W5 are boosted voltage buses; circuit breaker QFc plays a role in contacting the two groups of buses and is called a bus-tie circuit breaker (BTCB); incoming and outgoing lines of each circuit are equipped with circuit breakers and disconnectors, and the disconnectors on the bus side of the circuit breaker are called bus disconnectors, and those on the line side of the breaker are called line disconnectors; power from generators G1 and G2 is sent to the 10 kV bus, part of the power is supplied to the near-area loads by the cable lines, and the remaining power is sent to the boosted voltage buses W4 and W5 through step-up transformers T1 and T2; each cable feeder is fitted with reactor L to limit the short-circuit currents; since G1 and G2 are sufficient to supply the loads in this area, generator G3 is no longer connected to the 10 kV buses but is connected separately to the transformer T3 as a generator-transformer unit to minimize short-circuit currents in the generator voltage buses and the feeder lines.

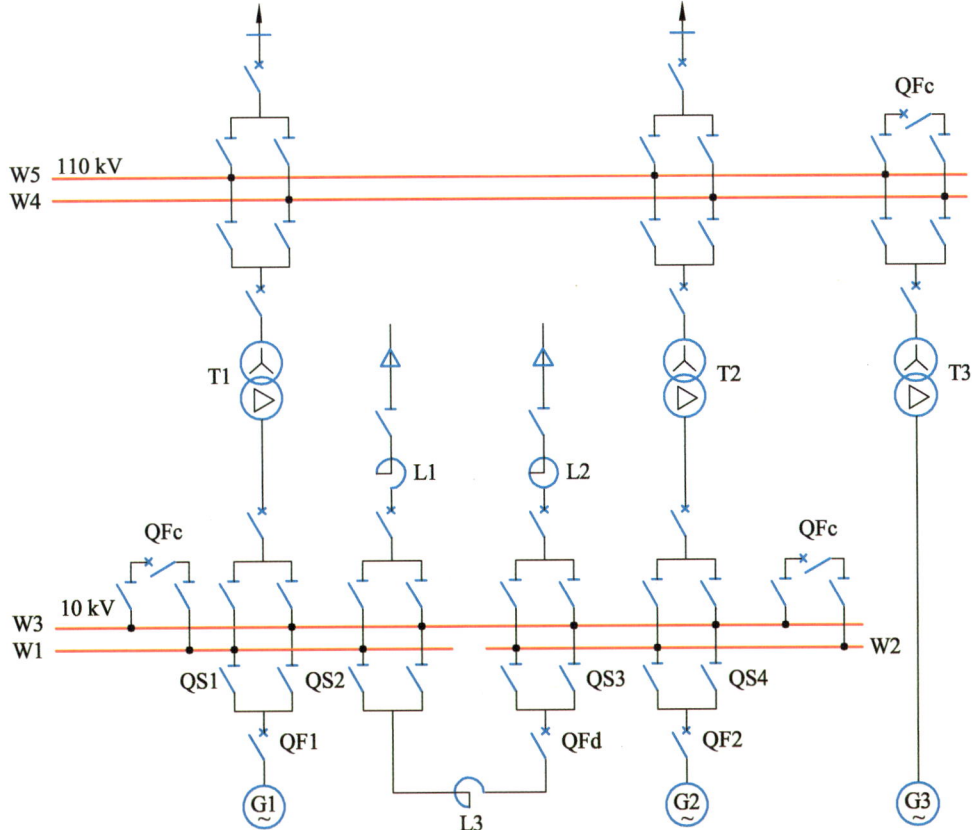

Fig. 1-16 Main Electrical Wiring Diagram of a Thermal Power Plant

The main electrical wiring of power plants and substations is selected as the optimal solution after design and technical and economic comparisons based on capacity, voltage class, load, and other conditions.

2. Power distribution unit

The electrical structure that receives and distributes electrical energy according to the main wiring diagram, consisting of a bus, switchgear, protective devices, measuring devices, and the necessary auxiliary equipment, is called a power distribution unit. Power distribution units are an important component of power plants and substations.

Power distribution units can be categorized into the following two types according to the installation location of the electrical equipment:

(1) In-house power distribution units. All equipment is installed inside.

(2) Outdoor power distribution units. All equipment is installed outside (i.e., open area).

Power distribution units can be categorized into the following two types according to the electrical equipment assembly method:

(1) Assembled power distribution units. Electrical equipment is assembled on site (inside or outside).

(2) Packaged power distribution units. The manufacturer pre-assembles the electrical equipment of each unit circuit in enclosed or unenclosed metal cabinets, which constitute the sub-divisions of the unit circuits. Most of the packaged power distribution units are of the in-house type, and there are also outdoor types.

Power distribution units can also be categorized in other ways, such as by voltage class, i.e., 10 kV power distribution units, 35 kV power distribution units, 110 kV power distribution units, 220 kV power distribution units, and 500 kV power distribution units.

项目二 变压器检修

变压器，特别是 110 kV 及以上的主变压器是升（降）压变电站的核心设备之一，如 550 kV 及以上的枢纽变电站降压后分配到下级 110 kV 或 220 kV 的区域变电站，再分配到 35 kV 或 10 kV 的终端站，同时 35 kV 及以下的配电变压器也是各级配电系统的核心设备之一，如生活中随处可见的厢式变电站、公变箱、台区变压器等均属于此类。由于变压器本身的功能决定其既在输电线路、变电站中起到核心作用，又能与用户侧紧密联系，在整个电网中的位置均处于各级电压的枢纽节点，因此，变压器的稳定运行一定程度上决定了电网的运行稳定。

由于 10 kV 变压器低压侧标称电压仅有 400 V，低压侧不属于高压范围，且其整体结构不超出 35 kV 及以上变压器结构范围，因此，现行的绝大多数标准、规范仅对 35 kV 及以上的变压器检修项目、质量提出要求。本章讲解内容适用于 35~220 kV 油浸式电力变压器。

模块一 变压器检修周期

变压器各类检修周期按照 2017 年《国家电网公司变电检修管理规定（试行）第 1 分册油浸式变压器（电抗器）检修细则》规定如下。

A 类检修周期按照设备状态评价决策进行。

B 类检修周期按照设备状态评价决策进行，应符合厂家说明书要求。

C 类检修现周期按如下规定进行：

（1）基准周期 35 kV 及以下 4 年、110（66）kV 及以上 3 年。

（2）可依据设备状态、地域环境、电网结构等特点，在基准周期的基础上酌情延长或缩短检修周期，调整后的检修周期一般不小于 1 年，也不大于基准周期的 2 倍。

（3）对于未开展带电检测设备，检修周期不大于基准周期的 1.4 倍；未开展带电检测老旧设备（大于 20 年运龄），检修周期不大于基准周期。

（4）110（66）kV 及以上新设备投运满 1~2 年，以及停运 6 个月以上重新投运前的设备，应进行检修。对核心部件或主体进行解体性检修后重新投运的设备，可参照新设备要求执行。

（5）现场备用设备应视同运行设备进行检修；备用设备投运前应进行检修。

（6）符合以下各项条件的设备，检修可以在周期调整后的基础上最多延迟 1 个年度：

① 巡视中未见可能危及该设备安全运行的任何异常。
② 带电检测（如有）显示设备状态良好。
③ 上次试验与其前次（或交接）试验结果相比无明显差异。
④ 上次检修以来，没有经受严重的不良工况。

D 类检修周期依据设备运行工况，及时安排，保证设备正常功能。

其中状态评价决策一般根据 2019 年《国家电网有限公司变电评价管理规定（试行）第 1 分册油浸式变压器（电抗器）精益化评价细则》《油浸式变压器（电抗器）状态评价导则》（DL/T 1685—2017）的要求依次对比评价标准、计算分值、评价设备状态等级、选择检修对策、编制检修项目，随后编制评价报告，在报告中即提出合理的检修周期。

Program 2　Transformer Maintenance

Transformers, especially main transformers of 110 kV and above, are one of the core equipment of step-up (step-down) substations. For example, a load-center substation of 550 kV and above is stepped down and distributed to the lower level 110 kV or 220 kV regional substations, and then to the 35 kV or 10 kV terminal stations. At the same time, 35 kV and below distribution transformer is also one of the core equipment at all levels of the distribution system. The box-type substation, utility transformers, and regional transformers, which can be seen everywhere in life, are all belong to this category. Due to the transformer's own function determining its core role both in the transmission lines and substations, but also the close contact to the user side, their positions in the entire power grid are at all levels of the voltage hub node, so the stable operation of the transformer to a certain extent determines the stability of the grid operation.

As the nominal voltage of the LV side of the 10 kV transformer is only 400 V, the LV side does not belong to the high-voltage range, and its overall structure does not exceed the structure of 35 kV transformers and above, therefore, most of the existing standards and codes only require the maintenance program and quality for transformers of 35 kV and above. The content of this chapter applies to 35—220 kV oil immersed power transformers.

Module 1　Transformer Maintenance Cycle

Transformer maintenance cycles for each category are stipulated as follows in accordance with the 2017 *Substation Maintenance Management Regulations of State Grid Corporation of China (Trial)—Volume 1: Detailed Rules for Maintenance of Oil Immersed Transformers (Reactors)*.

Class A maintenance cycle is carried out in accordance with the equipment condition evaluation decision-making.

Class B maintenance cycle is carried out in accordance with the equipment condition evaluation decision-making, and shall meet the requirements of the manufacturer's instructions.

Class C maintenance cycle is performed as follows:

(1) Reference cycle: 4 years for 35 kV and below, 3 years for 110 (66) kV and above.

(2) According to the state of equipment, geographical environment, power grid structure, and other characteristics, the maintenance cycle can be extended or shortened as appropriate on the basis of the reference cycle. Adjusted maintenance cycle is generally not less than 1 year, nor more than 2 times the reference cycle.

(3) For the equipment not carrying out live detection, the maintenance cycle is not greater than 1.4 times the reference cycle; for the old equipment (more than 20 years of age) not carrying out live detection, the maintenance cycle is not greater than the reference cycle.

(4) New equipment of 110 (66) kV and above that has been put into operation for 1 to 2 years and equipment that has been out of operation for more than 6 months before it is put into operation again shall be subject to maintenance. For equipment whose core components or main body have been deconstructed for maintenance and recommissioned, the requirements for new equipment may be applied.

(5) On-site standby equipment shall be subject to maintenance the same as operating equipment; standby equipment shall be subject to maintenance prior to operation.

(6) The maintenance can be delayed by up to 1 year on an adjusted cycle basis for equipment that meets all of the following conditions:

① No abnormalities that might jeopardize the safe operation of the equipment are observed during the inspection.

② The live detection (if any) shows that the equipment is in good condition.

③ No significant difference between the last test and its previous (or handover) test results.

④ Since the last maintenance, there is no serious adverse operating conditions.

Class D maintenance cycle is scheduled in time according to the operating conditions of the equipment to ensure the normal function of the equipment.

The condition evaluation decision-making is generally based on the requirements of the 2019 *Substation Evaluation Management Regulations of State Grid Corporation of China (Trial)—Volume 1: Detailed Rules for Lean Evaluation of Oil Immersed Transformers (Reactors)* and the *Guide for Condition Evaluation of Oil Immersed Transformers (Reactors)* (DL/T 1685—2017), which compare the evaluation criteria in order, calculate the scores, evaluate the condition level of the equipment, select the maintenance countermeasures, and prepare the maintenance items. Evaluation reports are subsequently prepared, in which reasonable maintenance cycles are proposed.

模块二 变压器部件和基础知识

一、部　件

　　油浸式电力变压器结构一般分为内部结构和外部结构两部分，内部结构决定了该变压器的功能、容量、主绝缘等核心性能，并通过一定的构架将之连接起来，形成一个整体，这个整体称为变压器的器身，包括铁心、绕组、引线及绝缘支架、油箱（外壳）这四部分（见图2-1、2-2）；除此之外所有的变压器部件都是依附于器身的核心功能进行安装，这些部件称为变压器附件。变压器附件形成一个变压器整体，这个整体也称为变压器本体，在变压器外部即可观察到的绝大多数部件都是附件（见图2-3至2-9）。

图 2-1　无绕组的三相铁心柱

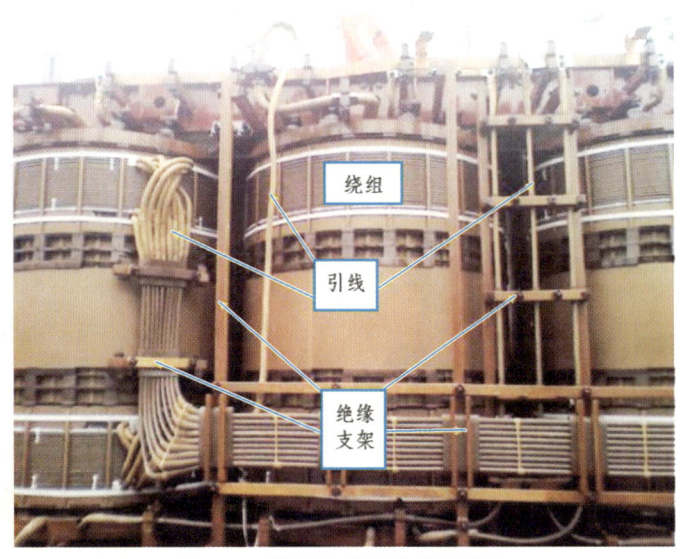

图 2-2　吊罩后的部分器身部件

1—油箱（外壳）；2—本体油枕；3—高压侧套管；4—低压侧套管；5—高压侧中性点套管；
6—低压侧中性点套管；7—散热器；8—分接开关机构箱；9—分接开关油枕；
10—高压侧套管升高座。

图 2-3 20 kV 油浸式变压器外观

图 2-4 变压器的温度表（从左至右分别
是油面温度表、上层油温温度表、绕组温度表）

图 2-5 油枕侧面的油位计

图 2-6 瓦斯继电器（气体继电器）

图 2-7 呼吸器

图 2-8　散热器底部的潜油泵（220 kV 及以上变压器）

1—分接开关；2—磁屏蔽块。

图 2-9　220 kV 变压器油箱内部

在上述结构的认识上，我们通常也按照各部分在检修时的功能整体性对其进行分类。

（1）套管及升高座检修包括：一般在低压侧的纯瓷充油套管，一般在高压侧的油纸电容型套管，升高座（套管型电流互感器）。

（2）储油柜及油保护装置检修包括：本体油枕（即储油柜），分接开关油枕，呼吸器（吸湿器）。

（3）分接开关检修包括：有载分接开关，无励磁分接开关（即无载分接开关）。

（4）冷却装置检修包括：散热器，强油循环冷却装置（含潜油泵，动力部分，管道），一般和潜油泵连接在一起的油流继电器，风机，冷却装置控制箱，水泵及喷淋泵。

（5）非电量保护装置检修包括：指针式油位计、气体继电器、温度计（表）、压力释放装置、突发压力继电器。

（6）二次端子箱检修。

（7）器身检修包括：绕组、铁心、引线及绝缘支架、油箱及管道、真空热油循环、吊装钟罩。

（8）排油和注油。

二、基础知识

1. 分类

根据电力变压器的用途和结构等特点可分为如下几类。

（1）按用途分：升压变压器（使电力从低压升为高压，然后经输电线路向远方输送）；降压变压器（使电力从高压降为低压，再由配电线路对近处或较近处负荷供电）。

（2）按相数分：单相变压器；三相变压器。

（3）按绕组分：单绕组变压器（为两级电压的自耦变压器）；双绕组变压器；三绕组变压器。

（4）按绕组材料分：铜线变压器；铝线变压器。

（5）按调压方式分：无载调压变压器；有载调压变压器。

（6）按冷却介质和冷却方式分：

① 油浸式变压器。冷却方式一般为自然冷却、风冷却（在散热器上安装风扇），强迫风冷却（在前者基础上还装有潜油泵，以促进油循环）。此外，大型变压器还有采用强迫油循环风冷却、强迫油循环水冷却等。

② 干式变压器。绕组置于气体中（空气或六氟化硫气体），或是浇注环氧树脂绝缘。它们大多在部分配电网内用作配电变压器。目前已可制造到 35 kV 级，其应用前景很广。

2. 容量等级

从电压等级上我们习惯称 220/380 V 为低压，3~35 kV 称为中压，110 kV 和 220 kV 称为高压，330 kV 和 500 kV 称为超高压，700 kV 和 1000 kV 称为特高压。

另外，变压器容量等级（单位：kV·A）主要包括：30、50、63、80、100、125、160、200、250、315、400、500、630、800、1000、1250、1600、2000、2500、3150、4000、5000、6300、8000、10 000、12 500、16 000、20 000、25 000、31 500、40 000、50 000、63 000、90 000、120 000、150 000、180 000、260 000、360 000、400 000。

工作中不能简单地按照电压等级来描述"大型"或"小型"变压器，比如在 10 kV 和 35 kV 等级变压器中都存在 1600 kV·A 这个容量。大小型变压器的应按容量划分，容量为 630 kV·A 以下的变压器统称为小型变压器；800~6300 kV·A 的变压器为中型变压器。8 000~63 000 kV·A 的变压器为大型变压器。90 000 kV·A 以上的变压器为特大型变压器。

3. 变压器铭牌和参数

图 2-10 中的变压器型号为 SFZ10-31500/110，变压器的型号分两部分：前部分由汉语拼音字母组成，代表变压器的类别、结构特征和用途；后部分由数字组成，表示产品的容量（kV·A）和高压绕组电压（KV）等级。完整的变压器型号由以下 11 个部分组成。

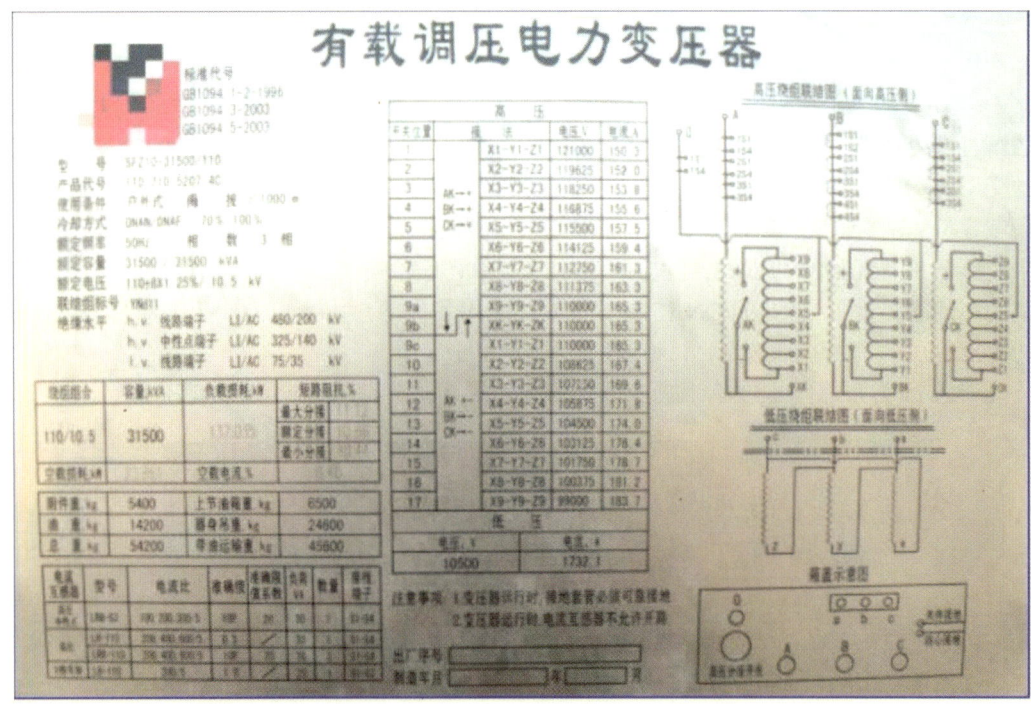

图 2-10　某品牌 110 kV 有载调压变压器铭牌

其含义分别如下：

第 1 格表示绕组数：S——三绕组；F——分裂绕组；双绕组不表示。

第 2 格表示相数：D——单相（或强迫导向）；S——三相。

第 3 格表示绝缘代号：C——成型固体；G——空气；油浸式不表示。

第 4 格表示冷却方式：J——油浸自冷；F——油浸风冷；FP——强迫油循环风冷；SP——强迫油循环水冷；G——干式空气自冷；C——干式浇注绝缘（环氧浇注）；自然风冷不表示。

第 5 格表示绕组材质或绕制形式：L——铝；铜不表示。

第 6 格表示调压代号：Z——有载调压；无励磁（无载）调压不表示。

第 7 格表示绕组绕制形式（一般是干式变压器）：B——箔式绕组；R——缠绕式绕组。

第 8 格表示设计序号或写作下标。

第 9 格表示额定容量。

第 10 格表示高压侧额定电压。

第 11 格表示其他使用要求：TH——湿热带（防护类型代号）；TA——干热带（防护类型代号）；GY——高原。

据此，SFZ10-31500/110 的含义为双绕组（不表示）三相（S）油浸（不表示）风冷（F）铜绕组（不表示）有载调压（Z）设计序号 10 型的变压器，额定容量为 31 500 kV·A，高压侧额定电压为 110 kV。

根据《变压器选用导则》要求变压器在所有情况下都应在铭牌上给出的项目有 11

类：变压器名称（如电力变压器、自耦变压器、有载调压变压器等），型号，产品代号；标准代号；制造厂名（包括国名）；出厂序号；制造年月；相数；额定容量（kV·A 或 MV·A）；额定频率（Hz）；各绕组的额定电压（V 或 kV）；各绕组的额定电流（A 对三绕组自耦变压器，还应注出公共线圈中长期允许的值）；联结组标号（6300 kV·A 以下的变压器，可不画联结示意图）；额定电流下的阻抗电压（实测值。如果需要，应给出参考容量，对多绕组变压器应表示出相当于 100%额定容量时的阻抗电压）；冷却方式（如果变压器具有几种冷却方式，除应表示出冷却方式外，还应以额定容量百分数表示出相应的冷却容量，如 ONAN/ONAF70/100%）；使用条件（户内、户外使用，超过 1000 m 的海拔等）；总质量（kg 或 t）；绝缘油质量（kg 或 t）。

图 2-11 为某品牌 10 kV 干式变压器铭牌，图 2-12 为某品牌 10 kV 变压器铭牌。

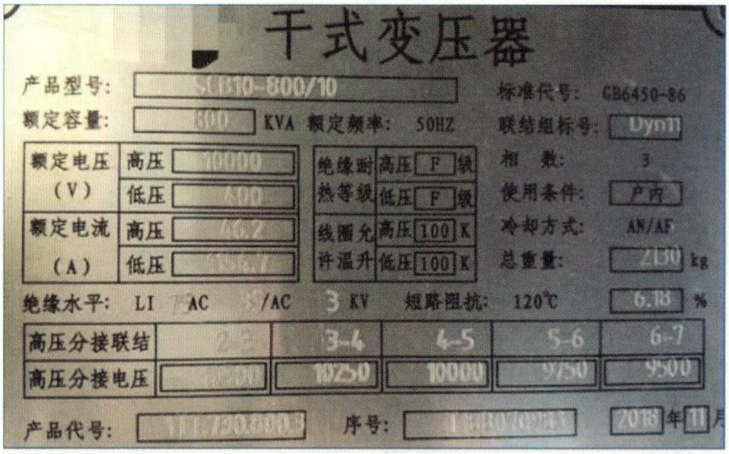

图 2-11 某品牌 10 kV 干式变压器铭牌

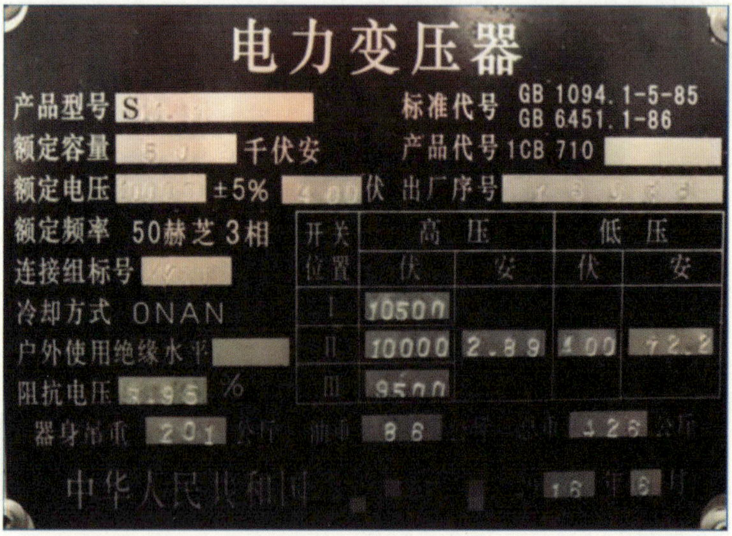

图 2-12 某品牌 10 kV 变压器铭牌

Module 2 Transformer Components and Basics

2.2.1 Components

The structure of an oil immersed power transformer is generally divided into internal and external parts. The internal structure determines the function, capacity, major insulation, and other core properties of the transformer, and connects them to form a whole through a certain framework. It is called the transformer body, including the four parts of the core, windings, leads and insulating support, and tank (housing), as shown in Figs. 2-1, 2-2. In addition, all the transformer components are attached to the body for installation on the basis of the core function, and these components are called transformer accessories. The transformer accessories form a whole transformer, which is also called the transformer body. The majority of the components that can be observed on the outside of the transformer are accessories, as shown in Figs. 2-3 to 2-9.

Fig. 2-1 Three-Phase Core Leg without Windings

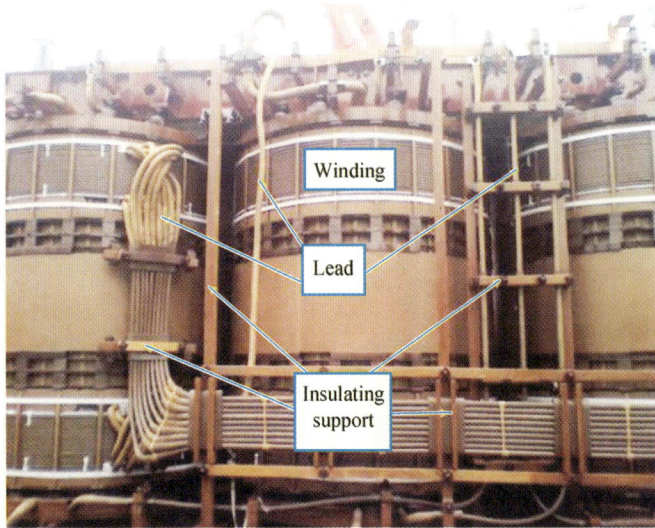

Fig. 2-2 Some of the Transformer Body Components after Bell Hood Hoisting

1—Oil tank (housing); 2—Body conservator; 3—HV side bushing; 4—LV side bushing; 5—HV side neutral point bushing; 6—LV side neutral point bushing; 7—Radiator; 8—Tap changer mechanism box; 9—Tap changer conservator; 10—HV side bushing turret.

Fig. 2-3　Appearance of 220 kV Oil Immersed Transformer

Fig. 2-4　Transformer Thermometers (From Left to Right: Oil Level Thermometer, Upper Oil Temperature Thermometer, Winding Thermometer)

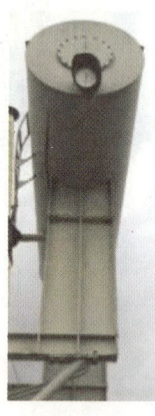

2-5　Oil Level Indicator on the Side of Conservator

Fig. 2-6　Gas Relay

Fig. 2-7　Breather

Fig. 2-8 Oil Submerged Pump at the Bottom of Radiator (220 kV Transformer and Above)

1—On-load tap changer; 2—Magnetic shielding block.
Fig. 2-9 Interior of 220 kV Transformer Tank

In recognition of the above structure, we usually also classify each part according to its functional integrity during maintenance.

(1) Maintenance of bushing and turret: pure ceramic oil-filled bushing generally on the LV side, oiled paper-condenser type bushing generally on the HV side, and turret (bushing-type current transformer).

(2) Maintenance of oil conservator and oil protection devices: body conservator (i.e. oil conservator), tap changer conservator, breather (moisture absorber).

(3) Maintenance of tap changer: on-load tap changer, non-excited tap changer (i.e. no-load tap changer).

(4) Maintenance of cooling devices: radiators, forced oil circulation cooling devices (including oil submerged pumps, power section, piping), oil flow relays generally connected

to the oil submerged pumps, fans, control box of the cooling device, water pumps, and spray pumps.

(5) Maintenance of non-electrical protective devices: pointer-type oil level indicator, gas relay, thermometer (meter), pressure release device, burst pressure relay.

(6) Maintenance of secondary terminal box.

(7) Maintenance of the transformer body: windings, cores, leads and insulating supports, oil tanks and pipes, vacuum hot oil circulation, bell hood hoisting.

(8) Oil draining and filling.

2.2.2 Basics

1. Classification

According to the use and structure and other characteristics, power transformers can be divided into the following categories.

(1) By the use: step-up transformer (boosting power from low voltage to high voltage and then sending it to distant places via transmission lines); step-down transformer (reducing power from high voltage to low voltage, and then supplying power to near or closer loads via distribution lines).

(2) By the number of phases: single-phase transformer; three-phase transformer.

(3) By the winding: single-winding transformer (for two-stage voltage autotransformer); double-winding transformer; three-winding transformers.

(4) By the winding material: copper-coil transformer; aluminum winding transformer.

(5) By the regulating method: no-load tap changing transformer; on-load tap changing transformer.

(6) By the cooling medium and cooling method:

① oil immersed transformer. The cooling methods generally include natural cooling, air cooling (fan installed on the radiator), and forced air cooling (an oil submerged pump is also fitted on top of this to facilitate oil circulation). In addition, large transformers also use forced oil circulation wind cooling, and forced oil circulation water cooling.

② Dry-type transformer. The windings are placed in a gas (air or sulphur hexafluoride gas) or insulated with cast epoxy resin. They are mostly used as distribution transformers in part of the distribution network. They can now be manufactured up to 35 kV class, and their applications are very promising.

2. Capacity rating

In terms of voltage class, we are used to describing 220/380 V as low voltage, 3—35 kV as medium voltage, 110 kV and 220 kV as high voltage, 330 kV and 500 kV as ultra-high voltage, and 700 kV and 1,000 kV as extra-high voltage.

In addition, the transformer capacity rating mainly includes 30, 50, 63, 80, 100, 125, 160,

200, 250, 315, 400, 500, 630, 800, 1,000, 1,250, 1,600, 2,000, 2,500, 3,150, 4,000, 5,000, 6,300, 8,000, 10,000, 12,500, 16,000, 20,000, 25,000, 31,500, 40,000, 50,000, 63,000, 90,000, 120,000, 150,000, 180,000, 260,000, 360,000, and 400,000 kV·A.

In practice, we cannot simply describe "large" or "small" transformers according to their voltage class. For example, a capacity of 1,600 kV·A exists in transformers of 10 kV and 35 kV class. Large and small transformers should be divided by capacity: transformers with a capacity of 630 kV·A or less are collectively referred to as small transformers; transformers with a capacity of 800—6,300 kV·A are medium-sized transformers; transformers with a capacity of 8,000—63,000 kV·A are large-sized transformers; transformers with a capacity of more than 90,000 kV·A are mega-transformers.

3. Transformer nameplate and parameters

Fig. 2-10 shows the nameplate of a 110 kV on-load tap changing transformer of a brand.

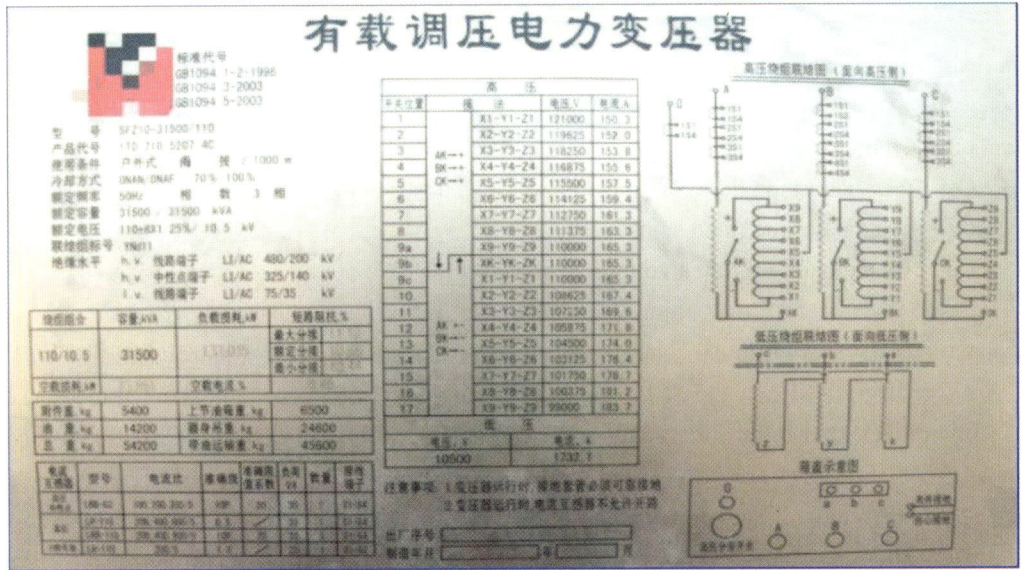

Fig. 2-10　Nameplate of a 110 kV On-Load Tap Changing Transformer of a Brand

Taking the transformer model SFZ10-31500/110 in the Fig. 2-10 as an example, the model of the transformer is divided into two parts: the first part is composed of Hanyu Pinyin letters, which represent the category, structural characteristics and use of the transformer, and the second part is composed of numbers, which indicate the capacity (KV·A) and the high-voltage winding voltage (KV) level of the product. The complete transformer model consists of the following 11 parts.

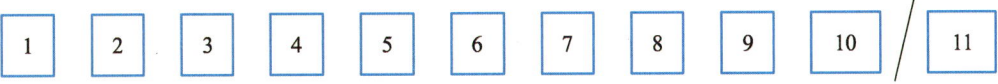

Their meanings are as follows:

Cell 1# indicates the number of windings: S—triple winding; F—split winding; double winding is not indicated.

Cell 2# indicates the number of phases: D—single-phase (or forced guided); S—three-phase.

Cell 3# indicates the insulation designation: C—molded solid; G—air; oil immersed type is not indicated.

Cell 4# indicates the cooling method: J—oil immersed self-cooled; F—oil immersed air-cooled; FP—forced oil circulation air-cooled; SP—forced oil circulation water-cooled; G—dry air self-cooled; C—dry cast insulation (epoxy casting); natural air-cooling is not indicated.

Cell 5# indicates the winding material or winding form: L—aluminum; copper is not indicated.

Cell 6# indicates the voltage regulation code: Z—on-load voltage regulation; un-excited (no-load) voltage regulation is not indicated.

Cell 7# indicates the winding form (generally dry-type transformer): B—foil winding; R—spiral wound winding.

Cell 8# indicates the design number or writing subscript.

Cell 9# indicates the rated capacity.

Cell 10# indicates the rated voltage on the HV side.

Cell 11# indicates other requirements for use: TH—wet tropical zone (protection type designation); TA—dry tropical zone (protection type designation); GY—plateau.

Accordingly, "SFZ10-31500/110" means a double-winding (not indicated), three-phase (S), oil immersed (not indicated), air-cooled (F), copper winding (not indicated), on-load voltage regulation (Z) transformer, with a design S/N of 10, a rated capacity of 31,500 kV·A, and a HV side rated voltage of 110 kV.

The *Guide for Choosing Power Transformers* requires that the transformer in all cases should be labeled on the nameplate with items of 11 categories: transformer name (such as power transformers, autotransformers, on-load tap changing transformers), model, product code; standard code; name of the manufacturer (including the name of the country); factory serial number; month/year of manufacture; number of phases; the rated capacity (kV·A or MV·A); rated frequency (Hz); rated voltage of each winding (V or kV); rated current of each winding (A; for three-winding autotransformer, long-term permissible values in the common coil should also be noted); tie-in module labeling (for transformers below 6,300 kV·A, the tie-in schematic diagram may not be drawn); impedance voltage under the rated current (measured value. If necessary, a reference capacity shall be given. For multi-winding transformers, the impedance voltage equivalent to 100% of the rated capacity should be indicated); cooling mode (if the transformer has several cooling modes, in addition to the cooling mode should be indicated, the corresponding cooling capacity should also be indicated as a percentage of the rated capacity, e.g., ONAN/ONAF70/100%); conditions of use (indoor and outdoor use, altitude of more than 1,000 m, etc.); total weight (kg or t); and insulating oil weight (kg or t).

Fig. 2-11 shows the nameplate of a 10 kV dry-type transformer, and Fig. 2-12 shows the nameplate of a 10 kV transformer.

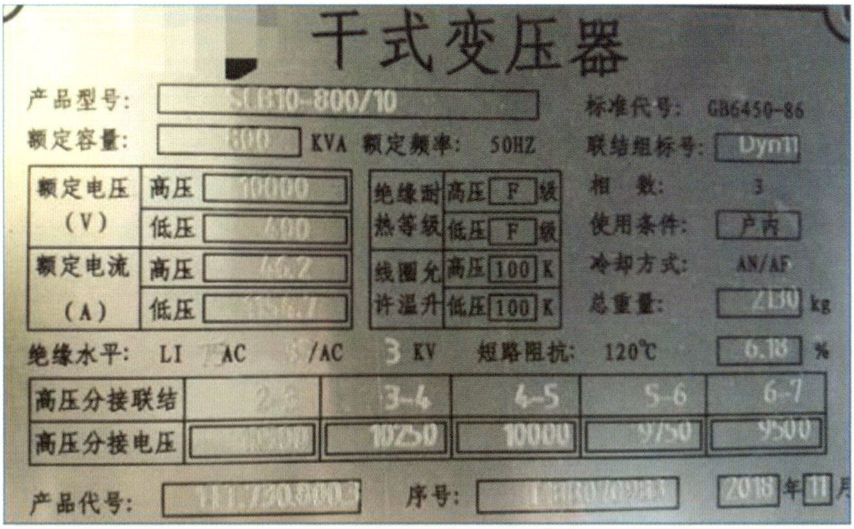

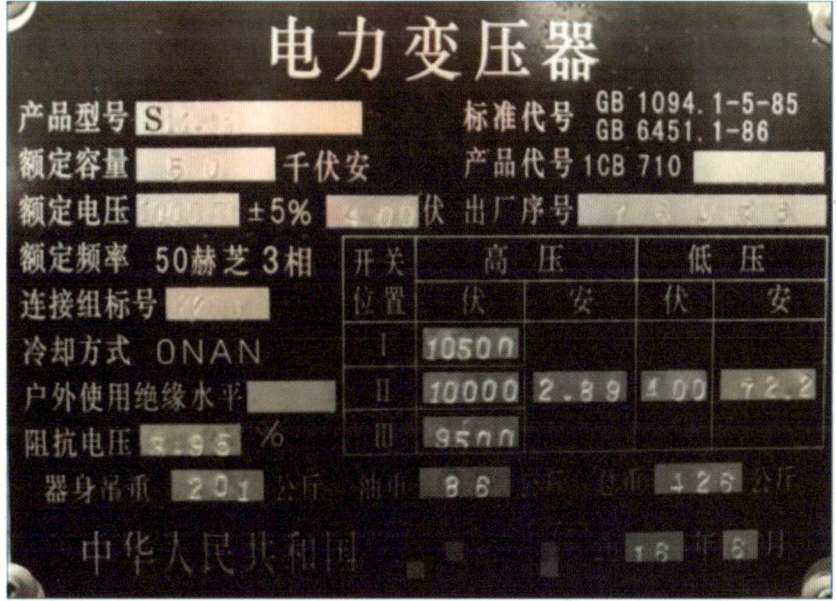

Fig. 2-11 Nameplate of a 10 kV Dry-Type Transformer of a Brand

Fig. 2-12 Nameplate of a 10 kV Transformer of a Brand

模块三　变压器绝缘

一、耐热等级

按照《电气绝缘的耐热性评定和分级》规定，电工产品绝缘材料按其耐热程度可分为常用的 7 个等级，它们的最高允许温度也各不相同。温度和绝缘材料通常对应关系见表 2-1。

表 2-1　耐热绝缘等级表

序号	耐热等级	温度/°C	常用材料	绕组温升限值
1	Y	90	棉纱、天然丝、再生纤维素为基础的纱织品，纤维素的纸、纸板、木质板等	非变压器用耐热等级
2	A	105	经耐温达标的液体绝缘材料浸渍过的棉纱、天然丝、再生纤维素等制成的纺织品、浸渍过的纸	60
3	E	120	聚酯薄膜及其纤维等	75
4	B	130	以云母片和粉云母纸为基础的材料	80
5	F	155	玻璃丝和石棉及以其为基础的层压制品	100
6	H	180	玻璃丝布和玻璃漆管浸以耐热的有机硅漆	125
7	C	>180	玻璃、电瓷、石英等	非变压器用耐热等级

由于习惯上的原因，无论对绝缘材料、绝缘结构和电工产品均笼统地使用"耐热等级"这一术语。但趋势是，对绝缘材料推荐采用"温度指数"和"相对温度指数"这两个术语；对绝缘结构则推荐采用"鉴别标志"这个术语；绝缘结构的"鉴别标志"只和所设计的特定产品发生联系；而对电工产品则保留采用"耐热等级"这个术语。

二、变压器绝缘分类

（一）绝缘分类

变压器的绝缘分为内绝缘和外绝缘两大类。内绝缘又分为主绝缘和纵绝缘两类。主绝缘是指线圈对它本身以外的其他结构部分的绝缘，包括它对油箱、铁心、夹件和压板的绝缘，对同一相内其他线圈的绝缘，以及对不同相线圈的绝缘（相间绝缘）。纵绝缘是指线圈本身内部的绝缘。它包括匝间绝缘、层间绝缘、线饼间的绝缘等。引线和分接开关的绝缘也可以同样划分。

变压器器身绝缘是主绝缘，是线圈-接地部分铁心和油箱的绝缘（主要是端部绝缘），线圈-其他线圈的绝缘（主要是同相线圈间主绝缘）。这种绝缘多为油-隔板和纸筒-油隙的形式。

表 2-2 线圈（绕组）内外绝缘位置表

分类	部件	绝缘类型	绝缘部位
内绝缘	线圈（绕组）	主绝缘	同相绕组之间
			异相绕组之间
			绕组对油箱
			绕组对铁心柱，绕组对旁柱之间
		纵绝缘	绕组端部对铁轭
			绕组线匝之间
			绕组饼间
	引线	主绝缘	绕组层间
			引线对地
		纵绝缘	引线对异相线圈
	开关	主绝缘	开关对地
			开关上不同绕组引线触头之间
		纵绝缘	同相绕组不同引线触头之间
外绝缘	套管		套管对各部接地之间
			异相套管之间

（二）绝缘材料

绝大多数变压器内部采用油、纸、纸板、纸板层压件等复合绝缘结构，来符合各种电压作用下的绝缘强度，即油纸绝缘结构。

变压器中油纸绝缘结构应用可以很大程度地提高绝缘强度，比油或绝缘纸单独使用时绝缘强度要高。

变压器中绝缘件在各部位（见图 2-13 至图 2-20）使用的作用如下。

（1）线圈间的绝缘纸筒：起主绝缘作用。

（2）线圈间内、外角环：起着加强线圈端部绝缘和改善端部电场作用。

（3）绝缘纸筒间的油道撑条。

（4）绝缘压板、绝缘垫板。

（5）静电环：改善端部电场作用。

（6）分相线圈间的绝缘隔板。

（7）引线绝缘夹木。

图 2-13　撑条隔开内外线圈

图 2-14　绝缘撑条

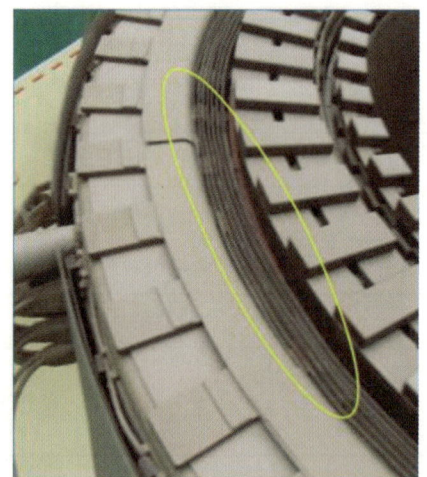
图 2-15　线圈绝缘纸筒

图 2-16　引线夹木

图 2-17　绝缘压板

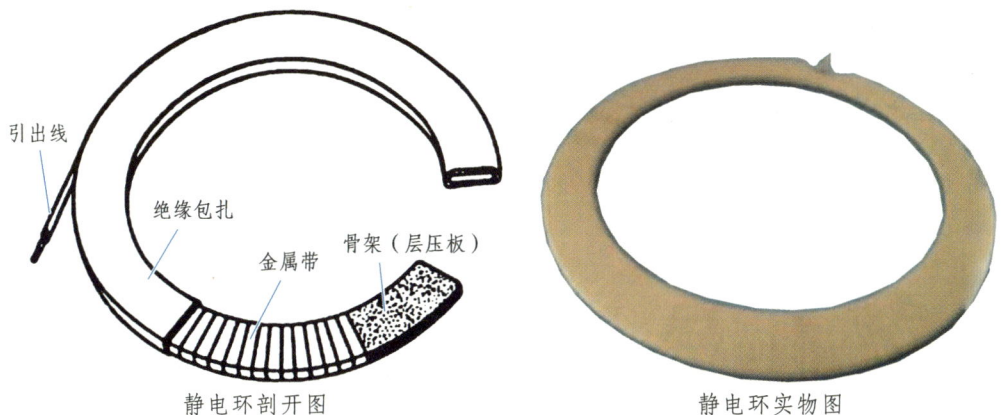

图 2-18 端部静电环

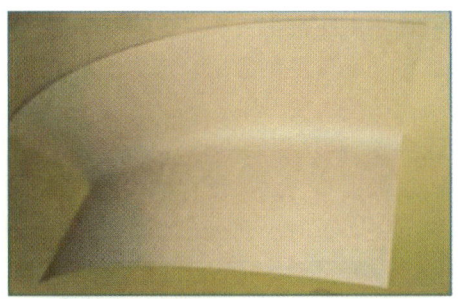

图 2-19 角环

图 2-20 绕组端部加强绝缘

(三)不均匀电场

1. 变压器内部稍不均匀电场

(1)同一铁心柱上不同电压等级线圈之间(除去绕组端部)。

(2)内绕组对有电屏蔽的铁心柱之间是同轴圆柱电场。

(3)不同相之间的圆柱电场。

2. 极不均匀电场

（1）内绕组对没有电屏蔽的铁心柱棱角处。
（2）引线对夹件等结构件。
（3）引线对油箱升高座的边缘。
（4）引线拐角对油箱壁。
（5）线圈端部对上下铁轭。
（6）线圈端部对绝缘压板的铁压钉等。

线圈的线饼之间、线饼中的线匝之间由于电极（导线）边缘之间圆角半径很小，会造成电场集中，故不能看作均匀电场。

3. 从手段上提高绝缘强度的途径

（1）采用端部加静电板（环）的办法来改善电极形状，见图2-20。
（2）端部增加角环，延长爬电距离，见图2-20。
（3）变压器引线绝缘加包一定厚度绝缘层，加大引线直径，电极表面光滑无毛刺，见图2-21。

超高压变压器采用出线装置，高压引线不得有金属裸露的地方（见图2-22、图2-23）。因为此时高压引线电极表面场强很高，如果没有绝缘覆盖，则该处的油中场强和它是同一个值，往往会引起局部放电（严重时出现火花放电），最终导致整个油隙击穿。

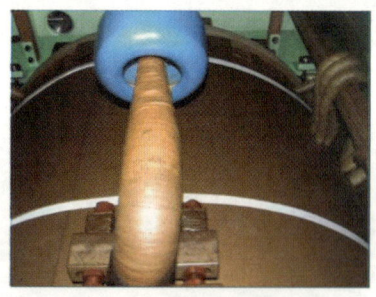

图2-21 加大引线直径

图2-22 出线端加装隔板（1，2，3）

图2-23 超高压变压器的出线装置

Module 3 Transformer Insulation

2.3.1 Temperature Classification

In accordance with the provisions of the *Thermal Evaluation and Classification of Electrical Insulation*, insulation materials for electrical products can be divided into seven commonly used grades according to their degree of heat resistance, and their maximum allowable temperatures are also different. The usual correspondence between temperature and insulating material is shown in Table 2-1.

Table 2-1 Temperature & Insulation Classification

S/N	Temperature classification	Temperature /°C	Common materials	Winding temperature rise limit
1	Y	90	Cotton yarn, natural silk, recycled cellulose-based textiles, cellulose-based paper, cardboard, and wood-based panels	Temperature classification for non-transformer use
2	A	105	Textiles made of cotton yarn, natural silk, and recycled cellulose impregnated with temperature-resistant liquid insulating materials, impregnated paper	60
3	E	120	Polyester film and its fiber	75
4	B	130	Materials based on mica flakes and mica paper	80
5	F	155	Glass wool and asbestos and laminated products based on them	100
6	H	180	Glass wool cloth and glass lacquer tubes impregnated with heat-resistant silicone lacquer	125
7	C	>180	Glass, electroceramics, quartz, etc.	Temperature classification for non-transformer use

For customary reasons, the term "temperature classification" is used generically for insulating materials, insulating structures, and electrical products. However, the trend is to use the terms "temperature index" and "relative temperature index" for insulating materials; the term "identification marker" for insulating structures; "identification marker" for insulating structures is only relevant for the particular product for which it is designed; while for electrical products, the term "temperature classification" is retained.

2.3.2 Transformer Insulation Classification

2.3.2.1 Insulation classification

Transformer insulation is divided into two categories: internal and external insulation, and internal insulation is divided into two categories: major insulation and minor insulation. The major insulation is the insulation of the coil from other parts of the structure other than itself, including its insulation from the tank, core, clamps, and pressboard, insulation from other coils within the same phase, and insulation from coils in different phases (interphase insulation). Minor insulation is the insulation within the coil itself. It includes turn-to-turn insulation, layer-to-layer insulation, and insulation between coil pancakes. The insulation of leads and tap changers can be similarly divided accordingly.

The transformer body insulation is the major insulation, including the insulation of the coil-grounded part of the core and the oil tank (mainly the end insulation), and the insulation of the coil-other coils (mainly the major insulation between coils of the same phase). This insulation is mostly in the form of oil-partition and paper tube-oil clearance.

Table 2-2   Positions of Coil (Winding) Internal and External Insulation

Classification	Components	Insulator type	Insulation part
Internal insulation	Coils (winding)	Major insulation	Between windings of the same phase
			Between windings of different phases
			Winding to the oil tank
			Between winding to the core leg, and winding to the side leg
		Minor insulation	Between winding turns
			Between winding pancakes
			Between winding layers
	Lead	Major insulation	Lead to the ground
			Leads to out-phase coils
		Minor insulation	Between different leads of a winding
	Switch	Major insulation	Switch to the ground
			Between contact terminals of different winding leads on the switch
		Minor insulation	Between contact terminals of different leads of the same phase winding
External insulation	Bushing		Between bushing to the grounding for each part
			Between bushings of different phases

2.3.2.2　Insulating materials

The interior of the vast majority of transformers is of oil, paper, cardboard, cardboard laminates and other composite insulating structures to comply with the insulating strength of various voltage effects, i.e., oiled paper insulating structures.

The application of oiled paper insulating construction in transformers can greatly increase the strength of insulation, which is higher than when oiled or insulating paper is used alone.

The roles of insulating parts (see Figs. 2-13 to 2-20) used in various parts of the transformer are as follows.

(1) Insulating paper tube between coils: the role of the major insulation.

(2) Inner and outer angle rings between coils: to strengthen coil end insulation and improve the end electric field.

(3) Strips for oil ducts between insulating paper tubes.

(4) Insulating pressboard and insulating padding plate.

(5) Electrostatic ring: improvement of end electric field.

(6) Insulating partition between split-phase coils.

(7) Lead insulating clamping wood.

Fig. 2-13　Inner and Outer Coils Separated by Strips

Fig. 2-14　Insulating Strips

Fig. 2-15　Coil Insulating Paper Tube

Fig. 2-16　Lead Clamping Wood

Fig. 2-17　Insulating Pressboard

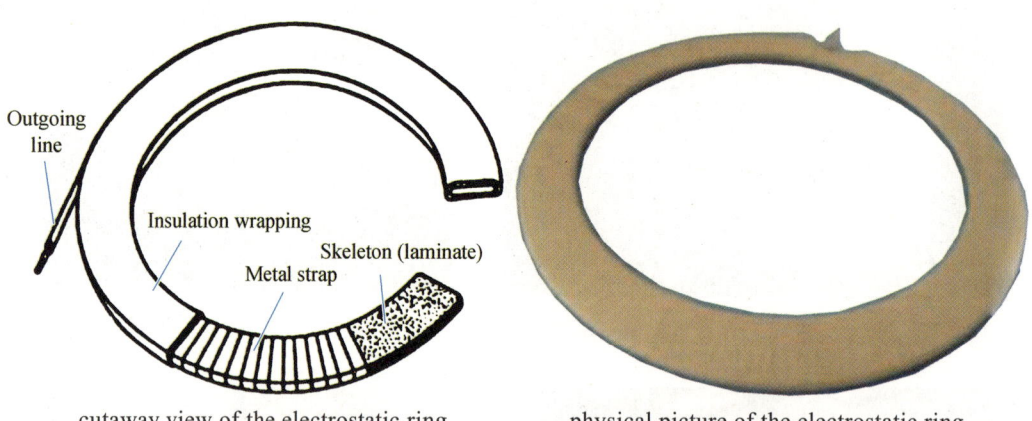

cutaway view of the electrostatic ring　　physical picture of the electrostatic ring

Fig. 2-18　End Electrostatic Ring

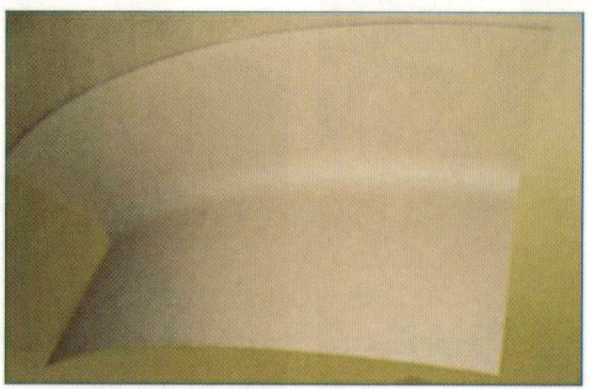

Fig. 2-19　Angle Ring

Fig. 2-20　Reinforced Insulation at Winding End

2.3.2.3　Nonuniform electric fields

1. Slightly nonuniform electric fields inside the transformer

(1) Between coils of different voltage classes on the same core leg (excluding winding ends).

(2) Coaxial cylindrical electric field between the inner windings to the electrically shielded core legs.

(3) Cylindrical electric fields between different phases.

2. Extremely nonuniform electric fields

(1) Edges and corners of the inner winding to a core leg not electrically shielded.

(2) Lead to structural members such as clamps.

(3) Edge of the lead to the turret of the tank.

(4) Corner of the lead to the wall of the tank.

(5) Coil end to the upper and lower yokes.

(6) Coil end to the iron pressure nails of the insulating pressboard.

Between the pancakes of the coils and between the turns of the pancakes cannot be regarded as a uniform electric field due to the small radius of the fillet between the edges of the electrodes (conductors), which causes a concentration of the electric field.

3. Ways to improve the insulating strength

(1) Improvement of electrode shape by adding electrostatic plate rings (loops) to the ends, see Fig. 2-20.

(2) Increasing the angle ring at the end to extend the creepage distance, see Fig. 2-20.

(3) Transformer lead insulation is wrapped with a certain thickness of insulation. The lead diameter is increased and the electrode surface is smooth and burr-free, see Fig. 2-21.

The UHV transformer is equipped with an outgoing device, and the HV leads shall not have a bare metal part (see Fig. 2-22, 2-23). Because at this time, the field strength of the HV lead electrode surface is very high, if there is no insulation coverage, the field strength in the oil at that location is the same value as it is, which tends to cause partial discharges (in severe cases, spark discharges), eventually leading to a breakdown of the entire oil clearance.

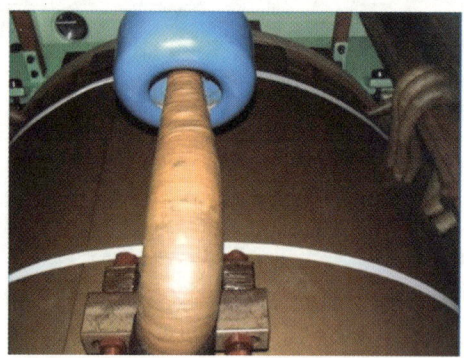

Fig. 2-21 Increased Lead Diameter

Fig. 2-22 Partitions (1, 2, 3) Added to the Leading-out Terminal

Fig. 2-23 Outgoing Device of UHV Transformer

模块四 变压器部件功能

1. 铁心

铁心是变压器的磁路部分。铁心是用导磁性能很好的硅钢片叠放组成的闭合磁路。

铁心同时又是安装线圈的骨架。变压器的原线圈和副线圈都绕在铁心上，对变压器电磁性能和机械强度是极为重要的部件。大多数变压器采用叠积式的铁心。对心式变压器来说，套装线圈的铁心柱总是由多级叠片组成一个近似圆形的截面，以求得在圆形线圈内部更有效地利用空间。

铁轭即铁心中不套线圈的部分一般可与心柱的截面形状相同，但有时为降低铁心高度采用变形轭。这时铁轭截面可做成矩形、椭圆形，再进一步要求降低铁心高度时，就要应用旁轭，旁轭截面形状一般均为椭圆形或矩形。

铁心具有不同的结构形式和用途，一般来说：

（1）单相双柱式铁心：适用于各种单相变压器。

（2）单相单柱旁轭式铁心：中间为一个芯柱，两边为旁轭，轭的截面为心柱截面的1/2。可降低上、下轭高，有助于减少附加损耗，适用于高压大容量单相电力变压器或大电流单相变压器。

（3）单相四柱式铁心（有两个旁轭）：中间为两个芯柱，两边为旁轭，可降低上、下轭高，有助于减少附加损耗，但电工钢片用量更多，体积大。有时在旁轭上安装调压和励磁线圈。适用于高压和超高压大容量单相电力变压器。

（4）三相三柱式铁心：在结构上与单相两柱式铁心是同一类型，只是多了一个芯柱，三柱线圈各自为一相引出。它是三相变压器最广泛应用的典型结构；

（5）三相三柱旁轭式铁心：中间为三个芯柱，各自为一相，两边旁轭和上、下端轭截面为芯柱截面的 1/3，主要是用来降低铁心的高度，便于运输。它适用于大容量三相电力变压器。铁心由硅钢片（晶合金、微晶合金）堆叠压制成整体，垂直于地面放置，为保证其稳定性，需要用夹紧装置是使之构成一个整体的紧固结构，这样的装置通常称为"夹件"，见图 2-24。夹紧装置在结构上应能可靠地压紧线圈、支撑引线、装置器身的绝缘件，并应具有器身在油箱中的定位结构，夹紧时的力应均匀，铁心片的边缘应不出现翘起，铁心片的接缝尽量要严合，在铁心励磁时噪声要尽量小。

1—铁轭；2—上夹件；3—钢拉带。

图 2-24　变压器铁轭位置夹件

铁心的绝缘当然属于变压器主绝缘。铁心的绝缘有两种，即铁心片间的绝缘以及铁心片与结构件件的绝缘：

铁心片间的绝缘是把心柱和铁轭的截面分成许多细条形的小截面，使磁通垂直通过这些小截面时，感应出的涡流很小，产生的涡流损耗也就很小。

铁心片间无绝缘时，磁通垂直通过的截面很大，感应的涡流大，截面厚度增加1倍，涡流损耗将增大至4倍。

铁心片间绝缘过小时，片间电导率增大，穿过片间绝缘的泄漏电流增大，将增加附加的介质损耗。

铁心片间绝缘过大时，铁心就不能认为是等电位的，必须把各片均连接起来接地，否则片间将出现放电现象，这是不方便的、不可取的。现在铁心用绝缘纸条做油道时，就需要把油道两侧的铁心片连接起来，然后由一个接地铜片引出（见图 2-25）。因此，铁心片间要有一定的绝缘，在标准测量方法下一般在 $60 \sim 105 \ \Omega/cm^2$。

冷轧工艺电工钢片的表面具有 $0.015 \sim 0.02$ mm 的无机磷化膜，可以满足这一要求，其他电工钢片则需要涂漆，检修时也需要涂漆，大型铁心有时要涂两遍漆。

由于铁心的金属结构件在线圈的电场作用下，具有不同的电位，与油箱电位有不同。虽然每层铁心之间电位差不大，也可能通过很小的绝缘距离而断续放电。放电一方面使油分解，加大发热，另一方面无法确认变压器在试验和运行中的状态是否正常。因此铁心及其金属结构件必须经油面接地，且要确保电气接通良好。这样的铁心接地我们称之为"铁心单点接地"，一旦发生铁心多点接地，形成局部放电或铁心的环流和涡流，会使铁心局部过热，使铁心损耗增加，甚至烧坏，而且过热造成的温升，使变压器油分解，产生的气体溶解于油中，引起变压器油性能下降，油中总烃大大超标，油中气体不断增加并析出(电弧放电故障时，气体析出量较之更高、更快)，可能导致气体继电器动作发信号，甚至使变压器跳闸（重瓦斯保护）。

图 2-25　铁心的接地引出线

图 2-26　铁心（未绕制绕组）

2. 绕组

绕组是变压器的电路部分。变压器有原线圈和副线圈，它们是用铜线或铝线绕成圆筒形的多层线圈，绕在铁心柱上，导线外边采用纸绝缘或纱包绝缘等（见图 2-26）。

对电力变压器来说，其线圈从套管的引出线通过母线与外部电网相连，当线路上发生单相接地故障或受雷电袭击发生大气过电压时，以及线路上断路器的动作产生操作冲击波时。线圈都将承受高于额定工作电压很多的瞬时过电压，当变压器线端发生短路故障时，线圈内部将有大于其额定电流许多倍的短路电流，这种短路电流会在线周内部产生巨大的机械力（轴向与辐向），同时会瞬时地使线圈温度升至极高的危险程度，由此产生了线圈的动热稳定问题。

因此，变压器线圈的质量和绕制方式不仅要考虑到变压器运行时各项额定工作参数，还必须充分考虑到遭受过电压时及短路故障时的绝缘强度及机械强度问题，线圈导线在瞬时短路电流通过时其温度上升，不能使其丧失机械强度和绝缘强度。

变压器的运行可靠性往往直接决定于线圈的结构设计和制造质量。根据电压等级的不同、容量的不同和使用条件的不同，将采用不同结构型式特点的线圈。这些特点是匝数、导线截面、并联导线换位、绕向、线圈连接方式和型式等。

高低压绕组的排列方式是由多种因素决定的。大多数变压器从绝缘方面考虑，是把低压绕组布置在高压绕组的里边。理论上，不管高压绕组或低压绕组怎样布置，都能起变压作用。但因为变压器的铁心是接地的，由于低压绕组靠近铁心，从绝缘角度容易做到。如果将高压绕组靠近铁心，则由于高压绕组电压很高，要达到绝缘要求，就需要很多的绝缘材料和较大的绝缘距离。这样不但增大了绕组的体积，而且浪费了绝缘材料。再者，由于大多数变压器的电压调节是靠改变高压绕组的抽头，即改变其匝数来实现的，因此把高压绕组安置在低压绕组的外边，引线也较容易（见图2-27）。

图 2-27　连续式绕组

油浸式变压器的绝缘基本上都是油浸纸和油的组合结构，这是将变压器绝缘油浸入纸或纸板中，使其介质常数有所下降不至于高出油的介电常数太多，这是油-纸绝缘的一个突出优点，其他材料如塑料、胶木等等油不能浸入的材料就没有这个特点。对高电压等级变压器，除绝缘外，要充分注意局部放电问题。为消除或降低局部放电量，应使电场分布尽量均匀避免局部畸变场强过高，除带电的金属部件要保持十分光滑无尖角毛刺外，绝缘零件也应做成无尖角毛刺，而且要保证油的充分浸透，没有残余气体或水份被包在绝缘材料内部。

3. 套管

套管是支持引出线之间及与变压器箱体的绝缘。110 kV 及以上套管主要采用全密封油浸纸绝缘电容式套管，套管自身密封不与变压器本体相通，并充有变压器油，法兰位置装设 CT 以供测量和保护用。

套管是变压器的载流组件，对变压器的绝缘性能有直接影响。是油浸式电力变压器箱外（附件）的主要绝缘装置，变压器绕组的引出线必须穿过套管，使引出线之间、引出线与变压器外壳之间绝缘，同时起到固定引出线的作用。

常用的套管型式见图 2-28：注油式、油纸电容式、胶纸式，另外还有油气式套管、电缆套管。

（1）注油式用于较低电压等级。

（2）油纸电容式用于高压与超高压。

（3）胶纸套管目前很少应用，宜用于均匀电场中，当用于不均匀电场中，绝缘性能不稳定。要保证局部放电量时一般不用胶纸套管。

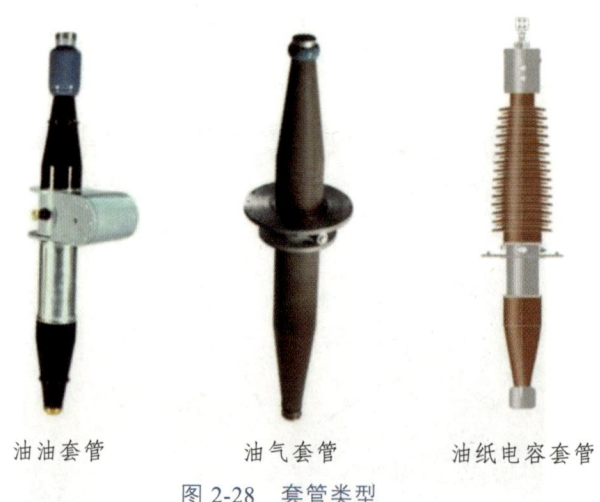

油油套管　　　油气套管　　　油纸电容套管

图 2-28　套管类型

在变压器运行中，长期通过负载电流，当变压器外部发生短路时通过短路电流。因此，对变压器套管有以下要求：

（1）必须具有规定的电气强度和足够的机械强度。

（2）必须具有良好的热稳定性，并能承受短路时的瞬间过热。

（3）外形小、质量小、密封性能好、通用性强和便于维修。

主绝缘主要由瓷套和油隙构成，属于不击穿、免维护部件。

套管结构见图 2-29，采用油纸电容芯子做主绝缘、穿缆式载流方式以及采用多组压力弹簧产生的轴向压紧力来套管实现的整体连接和主密封。252 kV 的套管在瓷件与连接套管连接处还辅心卡装结构以增强该部位的连接密封强度，套管的油枕、连接套筒均选用铸造铝合金，以减少磁滞和涡流损耗，降低发热及温升。连接套筒上设有试验和运行接地的抽头装置（末屏）及抽取套管内部油样的取油阀、释放变压器油箱顶部空气的放气塞。

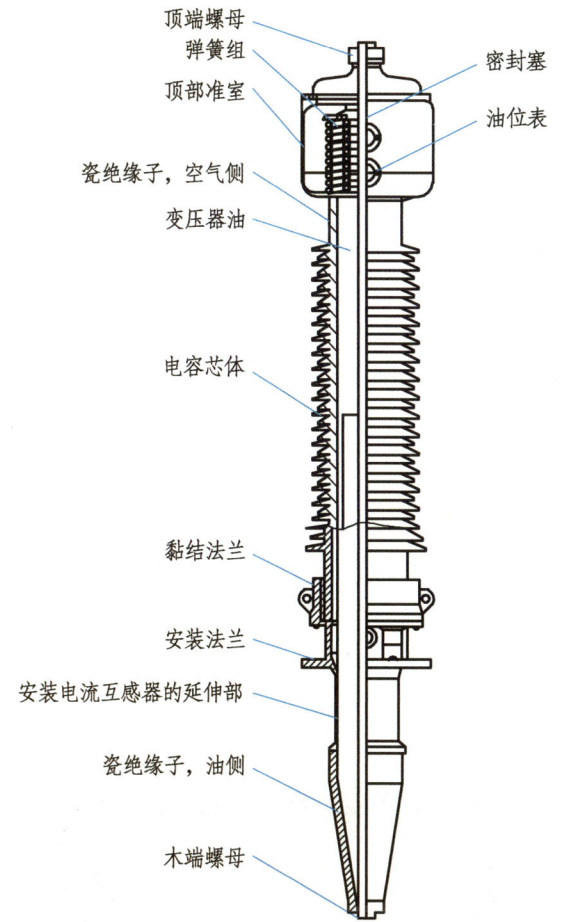

图 2-29 GOE2 型油纸电容套管结构

在运行中，由于套管表面瓷套长期暴露在空气中，可能有鸟粪、抛掷物、风吹物件等附着在上面，且在电场环境下，特别是重工业产业的地区，瓷套表面容易吸附脏污，这些都影响了放电爬距和干弧，因此套管的选择要结合周围环境，管户外端（或上端）外绝缘污秽等级代号及适用污区规定如下：

2——最小公称爬电比距为 20 mm/kV，中等污区。

3——最小公称爬电比距为 25 mm/kV，重污区。

4——最小公称爬电比距为 31 mm/kV，特重污区。

如：套管型号 BRDL1W-126/630-4，其中，BR——油纸电容式，D——短尾，L——加装电流互感器，W——防污型，最后 4 表示特重污区使用。若污秽等级不满足要求时，应喷涂防污闪涂料且状态良好或加装增爬裙且状态良好，现在一般喷涂的是有良好的憎水性和憎水迁移性的橡胶 RTV。

变电设备中几乎所有设备的绝缘子（瓷瓶、瓷套）检查都要求瓷套完好，无脏污、破损，无放电、电晕和电腐蚀现象，这当中"完好"的定义按照《国家电网公司变电评价管理规定（试行）第 1 分册油浸式变压器（电抗器）精益化评价细则》的要求，应该是绝缘子无碰损或开裂,单个缺釉不大于 25 mm²,釉面杂质总面不超过 100 mm²,另外套管油与变压器油是互相隔离的,因此其油位是单独观察,油位应该在观察窗口 2/3 至满油位。

4. 油枕

油枕保证油箱内充满油,使变压器缩少与空气的接触面,减少油的劣化速度;变压器油温随着负载和环境温度的变化而变化,当油的体积随着温度膨胀或缩少时,油枕起储油及补油作用;在油枕的侧面装有监视油位的油位计(玻璃式、连杆式、铁磁式)。胶囊式油枕内放置气囊,与呼吸器配合调节油位变化。

现可见的油枕形式有波纹式、胶囊式、隔膜式,隔膜式已逐渐被淘汰完,多数供电公司新购买的变压器都为主流的波纹式,现存的胶囊式也在有条件情况下通过技术改造成为波纹式。

以图 2-30 胶囊式油枕为例来看看其工作情况,为了使变压器油面与空气完全隔绝,其油位计间接显示油面。该油枕是通过在油枕下部的小胶囊,使之成为一个单独的油循环系统,当油枕的油面升高时,压迫小胶囊的油柱压力增大,小胶囊的体积被缩小了一些,于是在油位计反映出来的油位也高起来一些,且其高度与油枕中的油面成正比;相反,油枕中的油面降低时,压迫小胶囊的油柱压力也将减少,使小胶囊体积也相对地要增大一些,反应在油位计中的油面就要降低一些,且其高度与油枕中的油面成正比。换句话说,它使通过油枕油面的高、低变化,导致小胶囊压力大小发生变化,从而使油面间接地、间接保持油箱中的真空状态、成正比地反应油枕油面高低的变化。

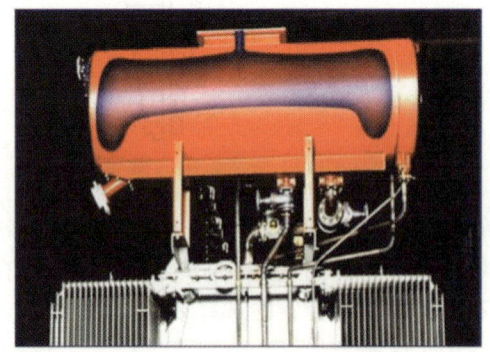

图 2-30 胶囊式油枕效果图

隔膜式和波纹式外形见图 2-31、图 2-32。

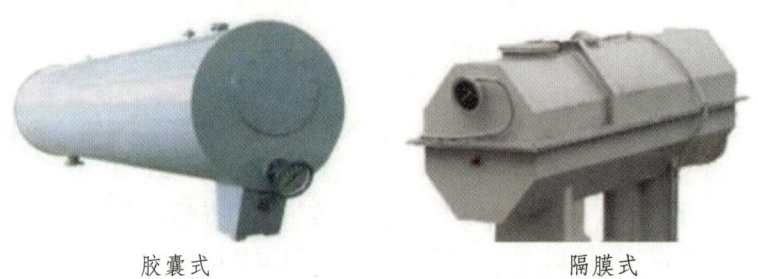

胶囊式 　　　　　　　隔膜式

图 2-31 胶囊式和波纹式外形

油枕外观　　　　　　　　　　　　部分截面

图 2-32　波纹式油枕

5. 油箱

油箱就是我们通常说的不带任何附件的变压器外壳，它承载铁心和线圈、变压器油。

6. 变压器油

用于绝缘、冷却、灭弧。

7. 调压装置

电压是电能质量的重要指标，电压波动的允许范围为 ±5%，超标后就要进行调节电压，调节方式现有两种。

（1）有载调压：变压器带负载状态下切换分接头位置。

（2）无载调压：变压器调压时不带任何负载，且与电网断开，在无励磁情况下变换绕组分接头。

不同的调压装置见图 2-33 至图 2-37。

1—有载分接开关；2—无载分接开关。

图 2-33　变压器上的调压装置

正反调压开关　　　　　　　多级粗细调压开关

图 2-34　M 型有载分接开关

外部接线头　　　　　　　　内部触头系统

图 2-35　V 型有载分接开关

无载分接开关　　　　　　　真空有载分接开关

图 2-36　其他分接开关

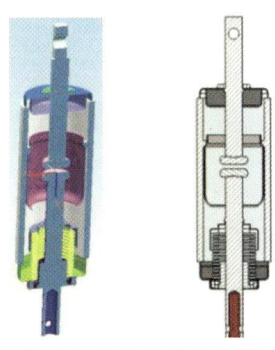 

触头在真空管中的位置　　　　真空管

图 2-37　真空有载分接开关的真空管

对于用户而言，有载调压更为友好，它提高了供电可靠率，这也是现在 35 kV 及以上电压等级变压器主流的调节方式，该调压方式下每次调节电压不超过前一个电压的 ±1.25%，总的调压范围是标称电压的 ±5% ~ ±10%。

有载分接开关调节的是变压器带负载改变绕组匝数比的装置，它主要用于电力系统中稳定负荷电压、联络电网、挖掘设备无功和有功出力以及调节负荷潮流，还用于工业所用的特种变压器上调节电压、电流和功率。按其切换时的过渡方式可分为电阻式和电感式，按结构形式可分为组合式 M 和复合式 V。

常用的 M 型分接开关最高工作电压为 40.5 ~ 252 kV，最大额定通过电流：三相为 300 ~ 600 A，单相为 300 ~ 1500 A。三相开关适用于 Y 形变压器中性点调压，单相开关适用于任意连接方式的变压器任意部位调压。根据变压器调压线圈结构设计需要，分接开关主要有线性调、多级线性调、正反调、粗细调、多级粗细调等调压方式。线性调最多调压级数为 18 级，正反调、粗细调为 35 级，多级线性调为 34 级、多级粗细调为 107 级。

在运行过程中，分接开关的通电触头始终是接通的，除了真空有载分接开关外，这些触头在持续运行、分合时的冷却、灭弧都是通过变压器油实现，有载分接开关有独立的油箱，无载分接开关却和变压器共用油箱，其他的分接开关的触头外层没有任何的与油隔离的部件。通常来说，这些触头只要处于运行中，就一直在"烧油"，造成分接开关对变压油的劣化影响非常快速，而快速劣化的油又反过来影响触头的接触效果和接触电阻情况，这样形成一个恶性循环，因此国家电网公司在多年以前就开始推广真空有载分接开关，但由于现有设备技改及开关本身成本的问题，一直在缓慢进行。

8. 净油器（在线滤油）

改善运行中绝缘油特性，防止绝缘油继续老化（多应用于有载调压油箱）。净油器内装吸附硅胶，吸收油中的水份及氧化物，使油保持洁净，延长油的使用年限，改善油的电气化学性能。

9. 冷却器

当变压器上层油温与下部油温产生温差时，通过散热器形成油的对流，经散热器冷却后流回油箱，起到降低变压器温度的作用。

不同的冷却形式也称为油循环方式，如自然风冷、强油风冷、导向风冷、水冷等，从本质上来讲就是加强循环速度或改变导热介质，增加油散热的效率。有的变压器为提高散热效率，在冷却器上加装了风机，这时是需要配备风机的控制装置，使其在一定温度标准下自动启动或停止，也叫冷却器控制箱。

油浸式变压器的散热过程是：先由热传导将铁心、绕组内部的热量传到其表面，然后传到油，再通过油的自然对流不断地将热量带到油箱、散热器油管的内壁，接着通过热传导把热量传到油箱、散热器油管的外表面，之后通过辐射和对流将热量散发到周围的空气中。

强迫油循环变压器的散热过程则是：用潜油泵将油上送入铁心中或绕组间的油道中，使其中的热量直接由具有一定流速的冷油带走，而变压器上层的热油用潜油泵抽出，经冷却器冷却后再送入变压器油箱底部，强迫变压器油进行油循环冷却。

结合第三节的变压器绝缘耐热等级，在运行发热状态下，变压器的上层油温一般不得超过 95 ℃，上层油温如果超过 95 ℃，变压器绕组的温度就要超过绕组绝缘物的耐热强度，从而加速绝缘物的老化。因此，为了使油不使过速氧化，电力行业不允许油温长期运行 95 ℃的极限温度，所以一般规定了 85 ℃这个上层油温的界限。

10. 呼吸器

当油枕内的空气随变压器油的体积膨胀或缩少时，排出或吸入的空气都经呼吸器，呼吸器内的干燥剂（硅胶）吸收空气中的水份，对空气起过滤作用，从而保证油的清洁。呼吸器内的硅胶变色过程：

$$蓝色 \to 淡紫色 \to 淡粉红（\geq 2/3 时需更换）$$

需要注意的是，呼吸器内的干燥剂吸湿剂不应自上而下变色，上部不应被油浸润，在市面上常见的有白色和蓝色两种，在这里使用的是蓝色，便于变色后分辨颜色。另外我们更换变色的干燥剂后，应罐装至顶部 1/6～1/5 处，而不是 2/3。

11. 气体继电器

有的旧规程规范中称为瓦斯继电器，是变压器的保护装置，一处装在变压器油箱至油枕的连接管上，一处装在有载分接开关油箱至其油室的连接管上，为便于调整位置，现在多数使用波纹管连接。

单浮子瓦斯继电器

双浮子瓦斯继电器

图 2-38　瓦斯继电器

（1）轻瓦斯（报警）：通过检测瓦斯继电器中积聚气体达到一定量时动作。

（2）重瓦斯（跳闸）：通过检测油流速度达到一定值时动作。

按照十八项反措和精益化评价要求，油灭弧有载分接开关应选用油流速动继电器，不应采用具有气体报警（轻瓦斯）功能的气体继电器；真空灭弧有载分接开关应选用具有油流速动、气体报警（轻瓦斯）功能的气体继电器。

220 kV 及以上变压器本体应采用双浮球并带挡板结构的气体继电器。

气体继电器安装时保持基本水平位置，气体继电器与储油柜之间连接的波纹管朝向油枕方向应有 1%~1.5% 的升高（油枕方向高），在不便于目测时，可以根据气体继电器与储油柜之间连接的波纹管，两端口同心偏差不应大于 10 mm 的标准来测量。以保证变压器内部故障时所产生的气体能顺利地上浮流向气体继电器。

12. 防爆（压力施放）阀

当变压器发生内部故障时，温度升高，油剧烈分解产生大量气体，使油箱内压力剧增，当压力达到防爆阀动作值时，防爆阀打开，油及气体油阀门喷出，防止变压器的油箱爆炸或变形（见图 2-39）。

图 2-39　压力施放装置

Module 4　Transformer Component Functions

2.4.1　Core

A core is the magnetic circuit part of the transformer. The core is a closed magnetic circuit composed of silicon steel sheets stacked with very good magnetic conductivity.

The core is also the skeleton of the installed coil. The primary and secondary coils of the transformer are wound on the core, which is an extremely important component for the electromagnetic properties and mechanical strength of the transformer. Most transformers use stacked cores. For core-type transformers, the core legs of the set coils are always made up of multiple stages of stacked pieces to form an approximately circular cross-section, to achieve more efficient use of space inside the circular coils. The yoke, i.e. the part other than the set coil, can generally have the same cross-sectional shape as the core leg. However, sometimes, in order to reduce the height of the core, a deformation yoke can be used.

At this time, the yoke cross-section can be made into a rectangular, or ellipsoidal. When further requirements to reduce the height of the core, we must use the side yoke. The cross-sectional shape of the side yoke is generally ellipsoidal or rectangular.

The core has different structural forms and uses, generally speaking:

(1) Single-phase, double-leg type core: It is suitable for all kinds of single-phase transformers.

(2) Single-phase, single-leg side yoke type core: It has a core leg in the middle, with side yokes on the two sides, and the cross-section of the yoke is 1/2 the cross-section of the core leg. It can lower the height of the upper and lower yokes, and help to reduce the additional loss, applicable to HV large-capacity single-phase power transformers or high-current single-phase transformers.

(3) Single-phase, four-leg core (with two side yokes): It has two core legs in the middle, with side yokes on the two sides, which can lower the height of the upper and lower yokes, and help to reduce the additional loss. However, the amount of electrical steel sheet is great, and the volume is large. Sometimes, it is necessary to install voltage regulating and excitation coils on the side yoke. It is suitable for HV and UHV large-capacity single-phase power transformers.

(4) Three-phase, three-leg core: It is of the same type in structure as the single-phase double-leg core, except that there is an additional core leg, and the three-leg coils are each led out for one phase. It is the most widely used typical three-phase transformer structure.

(5) Three-phase, three-leg side yoke type core: It has three core legs in the middle, each for a phase. The cross-section of the side yokes on both sides and the upper and lower end yokes account for 1/3 of the cross-section of the core leg. It is mainly used to lower the height of the core for easy transportation. It is suitable for large-capacity three-phase power transformers.

The core is made of silicon steel sheets (crystal alloys, microcrystalline alloys) stacked and pressed into a whole, placed perpendicular to the ground. In order to ensure their stability, clamping devices are required to make them into an integral fastening structure. Such a device is often referred to as a "clamp", see Fig. 2-24. The clamping device should be structured to reliably compress the coil, support the leads and insulating parts of the transformer body, and should have a structure for positioning the transformer body in the tank. The force during clamping shall be uniform. The edges of the core sheet shall be free from buckling. The jointing of the core sheet shall be as tight as possible. The noise during the core excitation should be as low as possible.

The insulation of the core is the major insulation of the transformer. There are two types of core insulation, i.e., insulation between core sheets and insulation between core sheets and structural members:

The insulation between the core sheets divides the cross-section of the core leg and yoke into many small, thin sections. When the magnetic flux passes perpendicularly through these small sections, the induced eddy current is very small, and the resulting eddy current loss is small.

1—Yoke; 2—Upper clamp; 3—Steel drawstring.
Fig. 2-24　Clamps on Transformer Yoke Positions

When there is no insulation between the core sheets, the magnetic flux that passes vertically through the cross-section is very large, and the induced eddy current is large. Every time the thickness of the cross-section increases by a factor of 1, the eddy current loss will increase to 4 times.

When the insulation between the core sheets is too small, the conductivity between the sheets increases, and the leakage current through the insulation between the sheets increases, which will increase the additional dielectric loss.

When the insulation between the core sheets is too large, the core cannot be considered equipotential. The sheets must be connected to the ground, otherwise, there will be a discharge between the sheets, which is inconvenient and undesirable. When insulating paper strips are used to make the oil duct of the core, it is necessary to connect the core sheets on both sides of the oil duct, and then led by a grounded copper sheet. Therefore, there should be certain insulation between the core sheets (see Fig. 2-5). It is typically in the range of 60 to 105 Ω/cm^2 under standard measurement methods.

Cold-rolled electrical steel sheet with 0.015—0.02 mm inorganic phosphate film on the surface can meet this requirement. Other electrical steel sheets need to be painted and subject to maintenance. Large cores are sometimes painted twice.

Fig. 2-25   Grounded Outgoing Line of Core

The metal structural members of the core have different potentials under the action of the electric field of the coil, which is different from the tank potential. Although the potential difference between each layer of the core is not large, it is possible to discharge intermittently through a small insulation distance. On the one hand, discharge decomposes the oil, and increases the heat; on the other hand, it is unable to confirm the normal condition of the transformer during testing and operation. Therefore, the core and its metal structural members must be grounded by the tank surface, and to ensure that the electrical connection is good. Such core grounding is called "core single-point grounding". Once the core multi-point grounding occurs, the formation of partial discharges or circulations and eddies in the core will make the core locally overheated, so that the core loss increases, or even burned out. And overheating caused by the temperature rise decomposes the transformer oil, resulting in gas dissolved in the oil, causing a decline in transformer oil performance. The total hydrocarbons in the oil greatly exceed the standard, and the gases in the oil are increasing and precipitating (the gas precipitation is higher and faster in the case of arc discharge faults). This may cause the gas relay to operate and signal, or even transformer trip (heavy gas protection).

2.4.2 Winding

A winding is the circuit part of the transformer. A transformer has a primary coil and a secondary coil, which are multilayered coils of copper or aluminum wire wound cylindrically around a core leg, with the outer edges of the wires subject to paper or cotton-covered insulation (see Fig. 2-26).

For power transformers, their coils are connected to the external grid through the bus via the outgoing lines of the bushing. When a single-phase grounding fault occurs on the line or an atmospheric overvoltage occurs from a lightning strike, and when the action of a circuit breaker on the line generates a switching impulse, the coil will be subject to a transient overvoltage much higher than the rated operating voltage. When a short-circuit fault occurs at the line end of the transformer, there will be a short-circuit current many times greater than its rated current inside the coil. This short-circuit current will produce huge mechanical force (axial and radial) inside the wire circumference, and at the same time, it will instantly make the coil temperature rise to a very high and dangerous level, which gives rise to the problem of dynamic thermal stability of the coil.

Fig. 2-26   Core (Unwound Winding)

Therefore, the quality of the transformer coil and winding method should not only take into account the rated operating parameters of the transformer operation, but also must take into full account the overvoltage and short-circuit fault insulation strength and mechanical strength problems. The coil conductor heats up during the passage of the transient short-circuit current. There must be no loss of mechanical or insulating strength.

The operational reliability of the transformer is often directly determined by the structural design and manufacturing quality of the coil. According to the different voltage classes, different capacities, and different conditions of use, coils with different structural type characteristics will be used. These characteristics include the number of turns, conductor cross-section, parallel wire transposition, winding direction, coil connection, and type.

HV and LV winding arrangement is determined by a variety of factors. However, in most transformers, the LV windings are arranged inside the HV windings. This is mainly from the insulation considerations. Theoretically, no matter how the HV or LV winding is arranged, it works as a transformer. However, since the transformer's core is grounded, keeping the LV windings close to the core is easily accomplished from insulation considerations. If the HV winding is close to the core, due to the high voltage of the HV winding, a lot of insulating material and a large insulating distance are needed to meet the insulation requirements. This not only increases the volume of the winding, but also a waste of insulation materials. Further, the voltage regulation of most transformers is accomplished by changing the taps, i.e., the number of turns, of the HV windings. It is therefore easier to place the HV winding on the outside of the LV winding and to lead the wires (see Fig. 2-27).

Fig. 2-27 Continuous Winding

The insulation of oil immersed transformers is basically a combined construction of oil-immersed paper and oil. This is where the transformer's insulating oil is immersed in paper or cardboard so that its dielectric constant is reduced but not too much above the dielectric constant of the oil. This is an outstanding advantage of oil-paper insulation. Other materials such as plastic, and bakelite that cannot be immersed in oil do not have this feature. In addition to insulation, due attention should be paid to the problem of partial discharge in HV transformers. In order to eliminate or reduce the amount of partial discharge, the electric field distribution should be made as uniform as possible to avoid too high local distortion field strength. In addition to the live metal parts should be kept very smooth and free of sharp corners or burrs, the insulating parts should also be made free of sharp corners or burrs. And to ensure that the oil is fully immersed, no residual gas or moisture is encapsulated within the insulating material.

2.4.3 Bushing

The bushings support the insulation between the outgoing lines and the transformer case. For 110 kV and above bushings, fully sealed oil-immersed paper insulated capacitive bushings are mainly used. The bushing itself is sealed from the transformer body and is filled with transformer oil. Its flange position is fitted with a CT for measurement and protection.

The bushing is the current-carrying component of the transformer and has a direct influence on the insulation performance of the transformer. It is the main insulating device outside the box (accessory) of the oil immersed power transformer. The leads of the transformer windings must pass through the bushing to insulate the outgoing lines from each other and from the transformer housing, fixing the outgoing lines at the same time.

Commonly used bushing types are shown in Fig. 2-28: oil-filled type, oiled paper-condenser type, adhesive paper type, as well as oil-gas type bushing, and cable bushing.

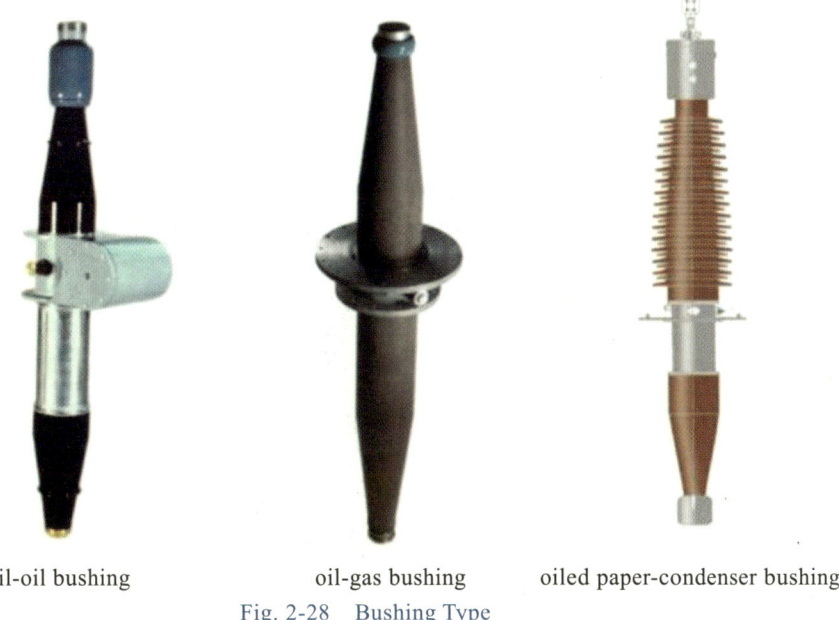

  oil-oil bushing   oil-gas bushing oiled paper-condenser bushing

Fig. 2-28 Bushing Type

(1) The oil-filled type used for lower voltage classes.

(2) The oiled paper-condenser type used for HV and UHV.

(3) The adhesive paper bushing is seldom used at present, and it is preferred to be used for a uniform electric field. When used in a nonuniform electric field, its insulation performance is not stable. To ensure the amount of partial discharge, the adhesive paper bushing is generally not used.

Transformers are passed by load currents for long periods of time during operation. When a short-circuit occurs outside the transformer, the short-circuit current is passed. Therefore, there are the following requirements for transformer bushings:

(1) It must have the specified electrical strength and sufficient mechanical strength.

(2) It must have good thermal stability and be able to withstand instantaneous overheating during a short circuit.

(3) Small shape, small quality, good sealing performance, great versatility, and easy maintenance.

The major insulation is mainly composed of porcelain bushing and oil clearance, which are no-breakdown and maintenance-free parts.

The structure of the bushing is shown in Fig. 2-29. The oiled paper condenser core is used as the major insulation, and the overall connection and major sealing are realized by the cable-trough current-carrying method and the axial compression force generated by multiple pressure springs. 252 kV bushing is also supplemented with a core cardboard structure in the connection between the porcelain parts and the connecting bushing to enhance the connection and sealing strength of the part. The conservator of the bushing and the connecting sleeve are made of cast aluminum alloy to reduce the hysteresis and eddy current loss, heat generation, and temperature rise. The connecting sleeve is equipped with a tapping device (end screen) for test and operation grounding and an oil valve for taking oil samples from inside the bushing, and a relief plug for releasing air from the top of the transformer tank.

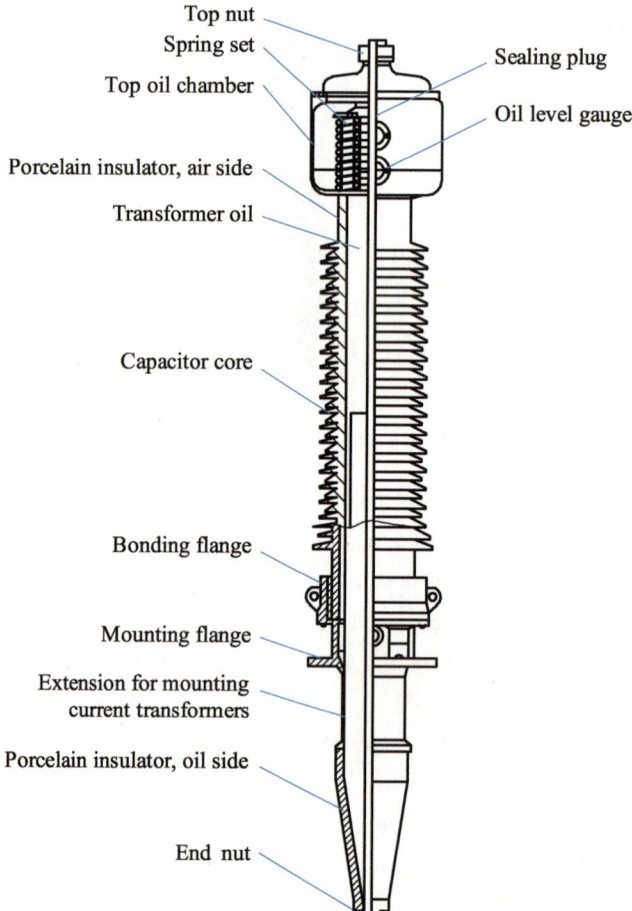

Fig. 2-29 Structure of GOE2 Oiled Paper Condenser Bushing

In operation, due to the long-term exposure of the surface of the porcelain bushing in the air, there may be bird droppings, thrown objects, or wind-blown objects attached to it. And in the electric field environment, especially in the area of heavy industrial industry, the surface of the porcelain bushing is easy to adsorb dirt. These all affect the discharge creepage and dry arcing distance. Therefore, the selection of bushing should be combined with the surrounding environment. The outer (or upper) end of the pipe has the following external insulation pollution grade code and applicable polluted area.

2—The minimum nominal creepage ratio distance of 20 mm/kV, moderately polluted area.

3—The minimum nominal creepage ratio distance of 25 mm/kV, heavily polluted area.

4—The minimum nominal creepage ratio distance of 31 mm/kV, extremely heavily polluted area.

Example: For bushing model BRDL1W-126/630-4, "BR" means oiled paper condenser type, "D" means short tail, "L" means current transformer mounted, "W" means anti-pollution type, and the last "4" means used in an extremely heavily polluted area. If the pollution grade does not meet the requirements, it should be sprayed with anti-pollution flashover coatings and in good condition, or retrofitted with an additional creepage extender and in good condition. Nowadays, it is generally sprayed with rubberized RTV that has good water-repellent and water-repellent migration properties.

In the substation equipment, almost all the insulators (porcelain insulators, bushings) of the equipment are required to be inspected with the bushings in intact condition, free of dirt, breakage, discharge, corona, or electro-corrosion. The definition of "intact" should be in line with the requirements of the *Substation Evaluation Management Regulations of State Grid Corporation of China (Trial)—Volume 1: Detailed Rules for Lean Evaluation of Oil Immersed Transformers (Reactors)*. Insulators should be free from touching or cracking, with a single glaze peel not more than 25 mm^2 and total glaze impurities not more than 100 mm^2. In addition, bushing oil and transformer oil are isolated from each other, so their oil levels are observed separately. The oil level should be 2/3 of the full oil level in the sight glass.

2.4.4 Conservator

A conservator is used to ensure that the oil tank is full of oil, so that the transformer reduces the contact surface with air and slows down the deterioration of oil. Transformer oil temperature varies with the load and ambient temperature changes. When the volume of oil expands or shrinks with the temperature, the conservator is used to store and replenish the oil. The side of the conservator is equipped with an oil level indicator to monitor the oil level (glass type, connecting rod type, ferromagnetic type). The capsule-type conservator is equipped with an air bladder inside, which is used in conjunction with a breather to regulate oil level changes.

Nowadays, the available forms of conservators include corrugated type, capsule type, and diaphragm type. The diaphragm type has been gradually eliminated. Most power companies are using the mainstream corrugated type for their new transformer purchased. The existing capsule type is also transformed into a corrugated type through technical modification when available.

Take Fig. 2-30 (capsule-type conservator) as an example to see how it works. In order to keep the oil level of the transformer completely isolated from air, its oil level indicator indirectly indicates the oil level. By means of small capsules in the lower part of the conservator, it is made into a separate oil circulation system. When the conservator's oil level rises, the pressure of the oil column compressing the capsule increased, and the volume of the capsule is reduced a little. So the oil level reflected in the oil level indicator is also higher, and its height is proportional to the oil level in the conservator. Instead, when the oil level in the conservator is lowered, the pressure of the oil column compressing the capsule will also be reduced, so that the capsule volume is relatively larger. The oil level reflected in the oil level indicator will be lower, and its height is proportional to the oil level in the conservator. In other words, the change in the oil level of the conservator leads to a change in the size of the pressure of the capsule, so that the oil level indirectly maintains the vacuum in the tank, proportionally reflecting the change in the conservator oil level.

Fig. 2-30 Rendering of Capsule-Type Conservator

See Figs. 2-31 and 2-32 for diaphragm type and corrugated type.

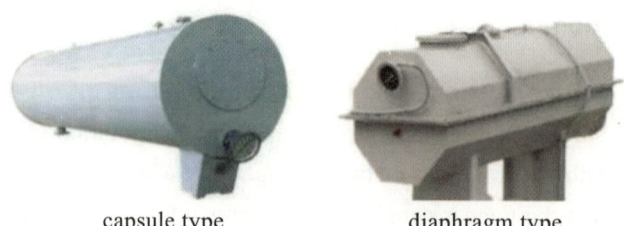

capsule type   diaphragm type

Fig. 2-31 Conservator

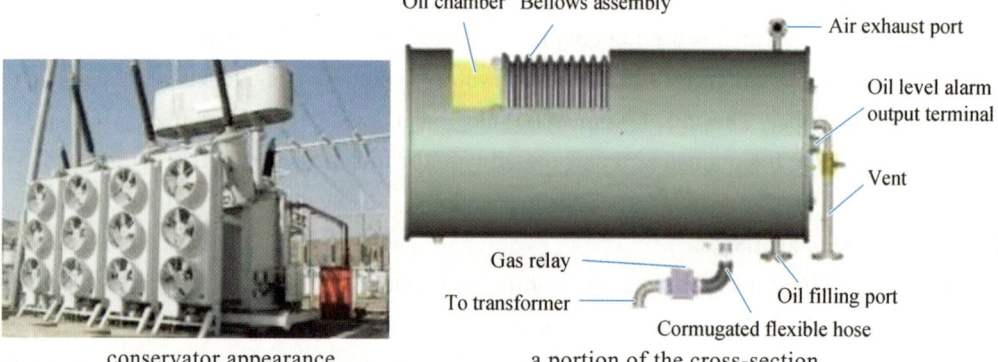

conservator appearance   a portion of the cross-section

Fig. 2-32 Corrugated Type Conservator

2.4.5 Oil tank

A tank is what we usually call the transformer housing without any accessories, which carries the core and coils and transformer oil.

2.4.6 Transformer oil

It is used for insulation, cooling, and arc extinguishing.

2.4.7 Voltage Regulator

Voltage is an important indicator of power quality. The permissible range of voltage fluctuation is ±5%, once exceeded, the voltage should be regulated. There are two types of regulation:

(1) On-load voltage regulation: Switching the tap position of the transformer under load;

(2) No-load voltage regulation: The transformer is regulated without any load and is disconnected from the grid, and the winding taps are changed under non-excited conditions.

The different voltage regulators are described in Figs. 2-33 to 2-37.

1—On-load tap changer; 2—No-load tap changer.

Fig. 2-33 Voltage Regulator on Transformer

For users, of course, on-load voltage regulation is more friendly and improves the reliability of the power supply. This is also the mainstream regulation method for transformers of 35 kV and above voltage class nowadays. In this regulation method, the voltage is adjusted each time by no more than ±1.25% of the previous voltage, and the total regulation range is ±5% to ±10% of the nominal voltage.

An on-load tap changer is a device for regulating a transformer with load by changing the turns ratio of the windings. It is mainly used in power systems to stabilize the load voltage, to contact the grid, to tap the reactive and active output of the equipment, and to regulate the load current, but also to regulate the voltage, current, and power on special transformers used in industry. According to its switching transition method, it can be divided into resistive and inductive types; according to the structural form, it can be divided into combined M and composite V types.

positive and negative tap changer multi-stage coarse and fine tap changer

图 2-34 M-Type On-Load Tap Changer

The maximum operating voltage of commonly used M-type on-load tap changers is 40.5—252 kV and the maximum rated let-through current is 300—600 A (three-phase) and 300—1,500 A (single-phase). The three-phase tap changers are suitable for regulating the neutral point of Y-connected transformers, while the single-phase tap changers are suitable for regulating any part of the transformer in any connection method. Depending on the design of the transformer's regulating coil structure, the main regulating methods of the on-load tap changer include linear regulation, multi-stage linear regulation, positive and negative regulation, coarse and fine regulation, and multi-stage coarse and fine regulation. The maximum number of regulating stages is 18 for linear regulation, 35 for positive and negative regulation and coarse and fine regulation, 34 for multi-stage linear regulation, and 107 for multi-stage coarse and fine regulation.

external terminal   internal contact terminal system

Fig. 2-35   V-Type On-Load Tap Changer

no-load tap changer   vacuum on-load tap changer

Fig. 2-36   Other Tap Changer

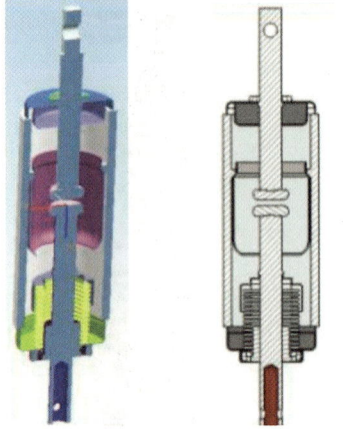 

position of the contact terminals in the vacuum tube　　　　vacuum tube

Fig. 2-37　Vacuum Tube for Vacuum On-Load Tap Changer

During operation, the energized contact terminals of the on-load tap changer are always connected. With the exception of the vacuum on-load tap changer, the cooling and arc extinguishing of these contact terminals during continuous operation, switching, and closing, is realized by means of transformer oil. The on-load tap changer has a separate oil tank, and the no-load tap changer and transformer share the same tank. Other on-load tap changers do not have any oil-isolating parts on the outer layer of the contact terminals. In general terms, these contact terminals "burn oil" as long as they are in operation. In this way, the on-load tap changer is affected by the rapid deterioration of the transformer oil, which in turn affects the contact effect and contact resistance of the contact terminals. This creates a vicious circle. The State Grid Corporation, therefore, started to promote vacuum on-load tap changers many years ago. However, this process has been slowed down due to technical changes in existing equipment and the cost of the tap changer itself.

2.4.8　Oil Purifier (On-line Oil Filtration)

It is used to improve the characteristics of the insulating oil in operation and to prevent the insulating oil from further aging (mostly used in on-load regulator tanks). The oil purifier is equipped with adsorbent silica gel to absorb water and oxides in the oil to keep the oil clean, prolong the service life of the oil and improve the electrochemical properties of the oil.

2.4.9　Cooler

When the temperature difference between the upper oil temperature of the transformer and the lower oil temperature occurs, the convection of oil is formed through the radiator and flows back to the tank after cooling by the radiator, which plays a role in reducing the temperature of the transformer.

Different forms of cooling are also known as oil circulation methods, such as natural air-cooled, forced oil air-cooled, guided air-cooled, and water-cooled. In essence, it is to strengthen the circulation speed or change the heat-conducting medium to increase the efficiency of oil heat dissipation. In order to improve the efficiency of heat dissipation, some transformers are equipped with fans on the cooler. At this time, it is necessary to equip the fan control device, so that it automatically starts or stops at a certain temperature standard. It is also called a cooler control box.

Heat dissipation process of oil immersed transformer: The heat inside the core and windings is first transferred to its surface by heat conduction and then to the oil. Through the natural convection of the oil, the heat is continuously brought to the inner wall of the oil tank and radiator oil hose, then to the outer surface of the oil tank and radiator oil hose by heat conduction, and then to the surrounding air by radiation and convection.

Heat dissipation process of forced oil circulation transformer: The oil submerged pump sends the oil up into the core or into the oil duct between the windings, so that the heat in it is directly taken away by the cold oil with a certain flow rate. The hot oil in the upper layer of the transformer is pumped out by the oil submerged pump, cooled by the cooler and then sent to the bottom of the transformer tank, forcing the transformer oil to carry out oil circulation cooling.

Combined with the transformer insulation and temperature classification in Section 3, in the operation of the heat state, the upper layer oil temperature of the transformer shall not exceed 95°C. If the upper layer oil temperature exceeds 95°C, the temperature of the transformer windings will exceed the heat-resistant strength of the winding insulation, thus accelerating the aging of the insulation. Therefore, in order to make the oil not make rapid oxidation, the power industry does not allow oil temperatures to run at the limit of 95°C for long periods of time, so the upper oil temperature limit of 85°C is generally specified.

2.4.10 Breather

When the air inside the conservator expands or shrinks with the volume of the transformer oil, the air discharged or inhaled goes through the breather, and the desiccant (silica gel) inside the breather absorbs the moisture in the air and filters the air, thus ensuring the cleanliness of the oil. The process of color change of the silicone inside the breather:

Blue → mauve → light pink (to be replaced when ≥2/3)

It should be noted that the desiccant and hygroscopic agent inside the breather should not be subject to color change from top to bottom, and the upper part should not be infiltrated by oil. There are two common ones on the market, white and blue. The blue one is used here to make it easy to distinguish the color after the color change. Also, after we replace the changed desiccant, it should be filled to 1/6—1/5 from the top, not 2/3.

2.4.11 Gas Relay

Some of the old rules and regulations refer to it as a gas relay. It is a protective device of the transformer, one is installed on the connecting pipe from the transformer tank to the conservator, and one is installed on the connecting pipe from the tank of the on-load tap changer to its oil chamber. In order to facilitate the adjustment of the position, most of them now use bellows for connection.

(1) Light gas (alarm): Actuated when a certain amount of gas is detected to have accumulated in the gas relay.

(2) Heavy gas (tripping): Actuated when the oil flow rate is detected to reach a certain value.

single float gas relay

double float gas relay

Fig. 2-38　Gas Relay

According to the 18 countermeasures and lean evaluation requirements, the oil arc extinguishing on-load tap changer should use an oil flow-rate relay, but not a gas relay with a gas alarm (light gas) function; The vacuum arc extinguishing on-load tap changer should use a gas relay with an oil flow-rate and gas alarm (light gas) function.

The body of the transformer of 220 kV and above should use a double-float gas relay with a baffle structure.

The gas relay should be installed in a basically horizontal position. The bellows connected between the gas relay and the oil conservator should have a rise of 1%—1.5% (height in the direction of the conservator) towards the conservator. When it is not convenient for visual inspection, it can be measured according to the bellows connected between the gas relay and the oil conservator, i.e. the concentric deviation of the two end ports should not be more than 10 mm, so as to ensure that the uplift of the gas generated during the internal failure of the transformer can flow smoothly to the gas relay.

2.4.12 Explosion-proof (Pressure Relief) Valve

When the transformer has an internal failure, the temperature rises, the oil decomposes violently and produces a large amount of gas, which increases the pressure in the oil tank dramatically. When the pressure reaches the operating value of the explosion-proof valve, the explosion-proof valve opens and the oil and gas are ejected from the valve to prevent the transformer tank from exploding or deforming (see Fig. 2-39).

Fig. 2-39　Pressure Relief Device

模块五　变压器专业巡视内容和评价标准

一、专业巡视内容

专业巡视是检修工作的一种，在不停电状态下进行，属于 D 类检修。按照《国家电网公司变电检修管理规定（试行）第 1 分册油浸式变压器（电抗器）检修细则》要求巡视项目如下。

（一）本体及储油柜

（1）顶层温度计、绕组温度计外观应完整，表盘密封良好，无进水、凝露，温度指示正常，并应与远方温度显示比较，相差不超过 5 ℃。

（2）油位计外观完整，密封良好，无进水、凝露，指示应符合油温油位标准曲线的要求。

（3）法兰、阀门、冷却装置、油箱、油管路等密封连接处应密封良好，无渗漏痕迹，油箱、升高座等焊接部位质量良好，无渗漏油。

（4）无异常振动声响。

（5）铁心、夹件外引接地应良好。

（6）油箱及外部螺栓等部位无异常发热。

（二）冷却装置

（1）散热器外观完好、无锈蚀、无渗漏油。
（2）阀门开启方向正确，油泵、油路等无渗漏，无掉漆及锈蚀。
（3）运行中的风扇和油泵、水泵运转平稳，转向正确，无异常声音和振动，油泵油流指示器密封良好，指示正确，无抖动现象。
（4）水冷却器压差继电器、压力表、温度表、流量表的指示正常，指针无抖动现象。
（5）冷却器无堵塞及气流不畅等情况。
（6）冷却塔外观完好，运行参数正常，各部件无锈蚀、管道无渗漏、阀门开启正确、电机运转正常。

（三）套管

（1）瓷套完好，无脏污、破损，无放电。
（2）防污闪涂料、复合绝缘套管伞裙、辅助伞裙无龟裂老化脱落。
（3）套管油位应清晰可见，观察窗玻璃清晰，油位指示在合格范围内。
（4）各密封处应无渗漏。
（5）套管及接头部位无异常发热。
（6）电容型套管末屏应接地可靠，密封良好，无渗漏油。

（四）吸湿器

（1）外观无破损，干燥剂变色部分不超过 2/3，不应自上而下变色。
（2）油杯的油位在油位线范围内，油质透明无浑浊，呼吸正常。
（3）免维护吸湿器应检查电源，检查排水孔畅通、加热器工作正常。

（五）分接开关

1. 无励磁分接开关

（1）密封良好，无渗漏油。
（2）挡位指示器清晰、指示正确。
（3）机械操作装置应无锈蚀。
（4）定位螺栓位置应正确。

2. 有载分接开关

（1）机构箱密封良好，无进水、凝露，控制元件及端子无烧蚀发热。
（2）挡位指示正确，指针在规定区域内，与远方挡位一致。
（3）指示灯显示正常，加热器投切及运行正常。
（4）开关密封部分、管道及其法兰无渗漏油。
（5）储油柜油位指示在合格范围内。
（6）户外变压器的油流控制（气体）继电器应密封良好，无集聚气体，户外变压器的防雨罩无脱落、偏斜。

（7）有载开关在线滤油装置无渗漏，压力表指示在标准压力以下，无异常噪声和振动；控制元件及端子无烧蚀发热，指示灯显示正常。

（8）冬季寒冷地区（温度持续保持零下）机构控制箱与分接开关连接处齿轮箱内应使用防冻润滑油并定期更换。

（六）气体继电器

（1）密封良好、无渗漏。
（2）防雨罩完好（适用于户外变压器）。
（3）集气盒无渗漏。
（4）视窗内应无气体(有载分接开关气体继电器除外)。
（5）接线盒电缆引出孔应封堵严密，出口电缆应设防水弯，电缆外护套最低点应设排水孔。

（七）压力释放装置

（1）外观完好、无渗漏，无喷油现象。
（2）导向装置固定良好，方向正确，导向喷口方向正确。

（八）突发压力继电器

外观完好、无渗漏。

（九）断流阀

（1）密封良好、无渗漏。
（2）控制手柄在运行位置。

（十）冷却装置控制箱和端子箱

（1）柜体接地应良好，密封、封堵良好，无进水、凝露。
（2）控制元件及端子无烧蚀过热。
（3）指示灯显示正常，投切温湿度控制器及加热器工作正常。
（4）电源具备自动投切功能、风机能正常切换。

（十一）干式变压器（干式铁心电抗器）

（1）设备外观完整无损，器身上无异物。
（2）绝缘支柱无破损、裂纹及爬电现象。
（3）温度指示器指示正确。
（4）无异常振动和声响。
（5）整体无异常过热部位，导体连接处无异常过热。

（6）风冷控制及风扇运转正常。

二、评价标准

在能进行以上专业化巡视的基础上，对每一项巡视内容进行精细的质量评判，如用什么标准判断"异常""破损"，有"异常""破损"又怎么判断能不能继续运行，这都关系到对缺陷的定性，直接影响检修策略的制定。按照《国家电网有限公司变电评价管理规定（试行）第 1 分册油浸式变压器（电抗器）精益化评价细则》中对运行中变压器可观察部件评价见本章后表 2-9。

Module 5 Content of Specialized Inspections and Evaluation Criteria for Transformers

2.5.1 Content of Specialized Inspections

Specialized inspection is a kind of maintenance work, which is carried out under non-power conditions and belongs to Class D maintenance. In accordance with the requirements of the *Substation Maintenance Management Regulations of State Grid Corporation of China (Trial)—Volume 1: Detailed Rules for Maintenance of Oil Immersed Transformers (Reactors)*, the inspection items are as follows.

1. Transformer body and oil conservator

(1) The appearance of the top-level thermometer and winding thermometer should be intact. The dials are well-sealed and free from water ingress and condensation. The temperature indication is normal and the difference is not more than 5°C when compared with the remote temperature display.

(2) The oil level indicator shall be intact in appearance, well-sealed, and free from water ingress and condensation. The indication should be in line with the requirements of the standard curve of oil temperature and oil level.

(3) Flanges, valves, cooling devices, oil tanks, oil piping, and other sealing connections should be well-sealed, with no signs of leakage. The oil tank, turret, and other welded parts are of good quality, with no oil leakage.

(4) No abnormal vibration or sound.

(5) The external grounding of the core and clamps should be proper.

(6) There is no abnormal heat in the oil tank and external bolts.

2. Cooling device

(1) Intact radiator appearance, with no corrosion or oil leakage.

(2) Correct opening direction of valves, with no leakage from oil pumps or oil lines, and no paint shedding or rusting.

(3) The running fan, oil pump, and water pump should run smoothly, and turn correctly, with no abnormal sound or vibration. The oil pump oil flow indicator is well sealed and indicates correctly without shaking.

(4) The differential pressure relay, pressure gauge, thermometer, and flow gauge of the water cooler are normal, and the pointer is not jittering.

(5) The cooler is not blocked and the airflow is smooth.

(6) The appearance of the cooling tower is intact, with normal operating parameters. There is no corrosion of parts and no leakage of pipes. The valves open correctly and the motors run normally.

3. Bushing

(1) The porcelain bushing is intact, without dirt, breakage, or discharge.

(2) The anti-pollution flashover coating, composite insulating bushing shed, and auxiliary bushing shed are not cracked, aging, or peeling.

(3) The oil level of the bushing should be clearly visible, the sight glass is clear, and the oil level indication is within the qualified scope.

(4) There should be no leakage at each sealing place.

(5) There is no abnormal heat in the bushing and joint parts.

(6) Capacitor-type bushing end screen should be grounded reliably, and sealed well, with no oil leakage.

4. Moisture absorber

(1) No appearance of damage. The color-changed part of the desiccant is not more than 2/3, and should not be changed from top to bottom.

(2) The oil level of the oil cup is within the oil level line. The oil is transparent without turbidity and breathes normally.

(3) The power supply of the maintenance-free moisture absorber should be checked, the smoothness of the drainage hole should be checked, and the heater should work normally.

5. Tap changer

(1) Non-excited on-load tap changer.

① Well sealed, with no oil leakage.

② The gear position indicator is clear and correct.

③ Mechanical operating devices should be free of corrosion.

④ Positioning bolts should be correctly positioned.

(2) On-load tap changer.

① The mechanism box is well-sealed and free from water ingress and condensation. The control components and terminals are not burnt or hot.

② The gear indication is correct, the pointer is in the specified area and is consistent with the remote gear.

③ The indicator light display is normal, and heater switching and operation are normal.

④ There is no oil leakage from the sealing part of the switch, pipe, and its flange.

⑤ The oil level of the oil conservator is within the qualified range.

⑥ The oil-flow control (gas) relay of the outdoor transformer should be well-sealed without gas collection. The rain cover of the outdoor transformer is not detached or skewed.

⑦ The on-line oil filtering device of the on-load tap changer has no leakage. The pressure gauge indicates below the standard pressure without abnormal noise or vibration. The control components and terminals are not burnt or hot, and the indicator light display is normal.

⑧ Anti-freezing lubricant shall be used in the gearbox at the connection between the mechanism control box and the on-load tap changer in cold winter areas (where the temperature is continuously below zero) and shall be replaced at regular intervals.

6. Gas relay

(1) Well sealed, no leakage.

(2) The rain cover is intact (for outdoor transformers).

(3) No leakage from the gas collector box.

(4) No gas in the window (except for the on-load tap changer gas relay).

(5) The cable exit holes of the junction box should be sealed tightly, the outgoing cable should be equipped with waterproof elbows, and the lowest point of the cable outer sheath should be equipped with drainage holes.

7. Pressure relief device

(1) Appearance is intact, with no leakage, or oil spraying.

(2) The guide device is well-fixed, the direction is correct, and the guide nozzle direction is correct.

8. Burst pressure relay

Appearance is intact, with no leakage.

9. Stop valve

(1) Well sealed, no leakage.

(2) The control handle is in the running position.

10. Control box and terminal box of the cooling device

(1) The grounding of the cabinet should be good, well-sealed, well-blocked, with no water, or condensation.

(2) The control components and terminals are not burnt or hot.

(3) The indicator light display is normal, and the switching temperature and humidity

controller and the heater work normally.

(4) The power supply has an automatic switching function, and the fan can be switched normally.

11. Dry-type transformer (dry-type core reactor)

(1) The appearance of the equipment is intact and undamaged, with no foreign matter on the transformer body.

(2) Insulated columns are free from damage, cracks, or creepage.

(3) The temperature indicator is correct in indication.

(4) No abnormal vibration or sound.

(5) There is no abnormal overheating in a whole, and no abnormal overheating in the conductor connection.

(6) Air-cooling control and fan operation are normal.

2.5.2 Evaluation Criteria

On the basis of the availability of the above specialized inspections, the quality of each inspection item should be finely judged, for example, what criteria are used to determine the "abnormality" or "breakage"; if there is an "abnormality" and "breakage", how to determine whether it can continue to operate. This is related to the characterization of defects, which directly affects the formulation of maintenance strategies. In accordance with the *Substation Evaluation Management Regulations of State Grid Corporation of China (Trial)—Volume 1: Detailed Rules for Lean Evaluation of Oil Immersed Transformers (Reactors)*, the evaluation of observable parts of the operating transformer is shown in Table 2-9 at the end of this chapter.

任务一 变压器吊罩后器身检查

一、工作任务

了解 110 kV 主变吊罩后对器身的检查内容，并能运用知识覆盖 35 kV 及以下的电压等级变压器器身检查。

二、引用标准

（1）《国家电网公司电力安全工作规程（变电部分）》（Q/GDW1799.1—2013）。

（2）国家电网公司生产技能人员职业能力培训专用教材——《变压器检修》。

（3）《国家电网公司变电检修管理规定（试行）第 1 分册油浸式变压器（电抗器）检修细则》。

（4）《国家电网公司变电运维管理规定（试行）第 1 分册油浸式变压器（电抗器）运维细则》。

（5）《电力变压器检修导则》（DL/T 573—2010）。

（6）《变压器分接开关运行维修导则》（DL/T 574—2010）。

三、现场气候及工作要求

（1）环境要求与变压器吊罩大修条件相同。

（2）分接开关选择器的绕组抽头连接已拆除，并已做好标记。

（3）分接开关心体已吊出。

（4）变压器钟罩已吊出并放置于多块承重枕木上，大盖边沿未与地面直接接触。

四、工作准备

1. 设备基本状况

该项工作是主变吊罩后的单项工作，器身（铁心、绕组、引线及绝缘支架、油箱）组成部件为静止状态，根据《电力变压器检修导则》要求检查工作"目测"是主要方式。正常工作中没有改变其运行、结构和接线方式的内容，主要是对指定对象的状态检查、清洗、测试。

2. 危险点及预控措施

（1）危险点——触电伤害。预控措施如下：

① 兆欧表测试线搭接需要使用绝缘手套。

② 绝缘手套需做检查。

③ 绝缘测试后记得放电。

（2）危险点——防止滑倒、坠落。预控措施如下：

① 工作人员应正确穿戴防滑鞋，戴鞋套。

② 器身上工作人员一旦落地，再次上梯前需重新更换新鞋套。

③ 在变压器上夹件上行走不得跑、跳。

（3）危险点——设备掉落。预控措施如下：

① 在变压器上夹件上工作所有个人用具和零件要放在工具包里。

② 用油泵吸油或冲洗时注意检查变压器底座是否有零件遗漏。

（4）危险点——人身伤害。预控措施如下：

清理附着在器身和变压器底座里的金属碎屑不得用手直接捧、抓，应戴新的棉手套或布手套，避免划伤。

（5）危险点——火灾。预控措施如下：

① 保证消防室内所有器材齐全，且有定期检查记录。

② 工作现场禁止吸烟，所有动火工作应远离油管、油管、滤油机等涉油区域，动火工作应有专门区域并有标示。

3. 工器具及材料选择

110 kV 主变吊心后器身检查的工器具及材料见表 2-3。

表 2-3　110 kV 主变吊心后器身检查的工器具及材料

类别	名称	规格型号	数量	备注
专用工具	安全帽		按人数配备	
	安全带		2 根	冲洗芯体,少量变压器油转移用
	油管与抱箍	250~500 mm 管径		配合油泵和滤油机
	电子式兆欧表			
	油罐			储油
	油泵	250~500 mm 管径	1 套	视油管的管径确定
	头顶灯(或手电)		按人数配备	可戴在安全帽上或挎背在身上
	梯子	3 m	1~2 把	爬上上夹件用
	接地线			电子式兆欧表外接接地用
个人工具	平口改刀			
	十字改刀			
	呆扳手	8-10,10-12,17-19		连接件
	活络扳手	150,300,600	各 1 把	
耗材	白绸布			擦拭心体及油室内部油渍
	棉纱			擦拭地面油渍
	砂纸	0 号或 1 号	各三张	锈蚀部分
	变压器油	25 号		冲洗心体用
	棉手套			
	布手套			
	雨衣	一次性,带兜帽		
	鞋套	一次性		
	酒精	工业酒精		清洗大盖法兰结合面

4. 作业人员分工

110 kV 主变吊心后器身检查工作人员分工见表 2-4。

表 2-4　110 kV 主变吊心后器身检查工作人员分工

序号	工作岗位	数量	工作职责
1	工作负责（监护人）	1	负责本次工作的人员分工、工作前的现场查看、作业方案制定、工作票的编制，办理工作许可手续，召开工作班前班后会，负责作业工程中的安全监督、工作中突发情况处理，工作质量监督，工作后的总结
2	操作人员	1	负责检修工作，结果核查，报表、记录填写
3	辅助作业人员	1	协助需要配合的工作，操作过程中的书面和照相记录，工作完成后现场清洁

5. 操作流程

器身整体检查及维护操作流程见表 2-5。

表 2-5　器身整体检查及维护操作流程

序号	作业内容	作业步骤及标准	安全和技术注意事项
1	前期准备工作	（1）分接开关选择器的绕组抽头已做好标记并已拆除。 （2）变压器钟罩已调出并放置于多块承重。枕木上大概边沿位与地面直接接触。 （3）冲洗用油管无打卷，尽量放直。 （4）底座放油孔已打开，便于冲洗，时有排出	（1）正确穿戴安全帽，工作服，劳保手套。 （2）严禁无关人员车辆进入作业现场
2	检查工器具及耗材	（1）将工器具按要求准备齐全并摆放整齐检查工器具外观、性能和试验合格证，无遗漏。 （2）对电子式兆欧表做通段检查显示报警放电功能均完好。 （3）一次性带兜帽雨衣，雨鞋套单独放置，未与其他脏污物品混用。 （4）梯子底脚的防滑胶垫完好	（1）工器具外观检查合格，无损伤、变形失灵现象，合格证在有效期内。 （2）一次性雨衣和鞋套要避免污染，以免穿戴后在气身上，工作对气身造成二次脏污
3	冲洗器身	（1）操作人员正确穿戴雨衣，用橡筋或白纱袋扎好袖口、裤脚。 （2）尽量站在干净油布上穿鞋套，随即登梯上变压器身。 （3）将冲洗用油管递交工作人员。 （4）油泵向油管供油，若有功率可调节，则待操作人员手持油管端已出油，再逐渐调大功率。 （5）操作人员沿器身夹件来回冲洗变压器铁心、绕组、夹件。 （6）用白绸布清洁上夹件及可触及的部分；	（1）上下楼梯时由于梯子搭在绕组上需要专人扶持。 （2）冲洗用油可用从变压器中放出的油。 （3）有功率可调节的油泵，避免一次性大功率出油，避免油管突然绷直导致操作人员手持不稳，掉落或伤及操作人员。

续表

序号	作业内容	作业步骤及标准	安全和技术注意事项
3	冲洗器身	（7）待操作人员下地后，其他工作人员在冲洗器身侧面	（4）由于冲洗后会刷下大量的脏污和金属碎屑，因此冲洗后的油应用单独油罐或油桶盛放，不再过滤使用。 （5）操作人员在器身上方冲洗时尤其应注意冲洗油道、夹件缝隙、上铁轭、绕组铁心。 （6）若操作人员已下地，再次上去时需要更换新鞋套
4	检查铁心	（1）检查铁心的外观完整程度，主要注意边沿层级是否均匀，叠片紧密程度，其表面和边沿应无翘起或弯折，用清洁白布清洁铁心表面的油污、铁屑等脏污，特别不能出现大范围的铁心层间缝隙，铁心搭接部分端头的弯曲短路。 （2）观察油道应排列整齐，无垫块脱落、堵塞、滑动等不稳定现象。 （3）铁心与上下夹件、压板间的绝缘应完好，不能出现积压、磨损导致的破裂。 （4）铁心绝缘层、夹件和其他金属件的防锈漆应无放电烧伤和过热痕迹。 （5）铁心组间、夹件、穿心螺栓、钢拉带（或环氧脂带）应有一点可靠接地。 （6）铁心接地片插入应牢靠，其外露部分应绝缘包扎，防止铁心短路。 （7）用兆欧表 2500 V 级以上挡位测量铁心及夹件对地绝缘	（1）禁止按照旧版变压器检修要求将卷曲铁心或不平的部分敲平，会损伤铁心表面绝缘。 （2）特别注意各金属件的尖端放电痕迹，另外螺母的边沿倒角部分也可能对夹件放电。 （3）铁心及夹件对地绝缘在《电力变压器检修导则》上要求是大于 1 MΩ，但"五通"要求 ≥ 100 MΩ（新投运 1000 MΩ）（750 kV 及以下），并按此实行
5	绕组检查	（1）（若有）则检查相间隔板和围屏无破损、变色、变形、放电痕迹。 （2）（无围屏）目测绕组应清洁、无油垢、无变形、无过热变色和放电痕迹。 （3）绕组整体无倾斜、位移，导线无明显弹出现象。 （4）油道经冲洗，应无油垢和其他杂质存积。 （5）绕组上下垫块、级间撑条无位移和松动，排列整齐，垫块外露出绕组长度不能超过绕组导线的厚度。 （6）用指压方式检查绕组绝缘状态	除正面观察绕组整体外，还应在侧面观察夹件整体无左右、上下错位，否则会引起绕组底部悬空，绕组垮塌，铁心与绕组短路等现象

项目二　变压器检修

续表

序号	作业内容	作业步骤及标准	安全和技术注意事项
6	引线及绝缘支架检查	目测检查引线，应无断股损伤、过热痕迹。用目测或手捏的方法，在绕组的引线引出头位置，绝缘夹板位置，焊接部位，分接开关引线接头等拉伸、弯曲应力集中部位检查其是否有应力集中现象。 （3）目测检查引线绝缘包扎情况，应完好，无变形起皱、变脆、破损、断股、变色情况。 （4）检查引线接头，接头表面应平整光滑，无毛刺，过热性变色现象，接头面积应大于其引线截面的3倍以上。 （5）检查穿缆套管的穿缆引线，其应用白纱带半叠包一层。 （6）在低压侧检查大电流引线是否包绝缘，未包扎的应按规定进行包扎。 （7）检查引线与各部件之间的绝缘距离。 （8）绝缘支架有无松动、损坏和位移，应无破损，裂纹，弯曲变形及烧伤现象。 （9）绝缘支架与铁夹件的固定情况，可用钢螺栓，绝缘件与绝缘支架的固定应用绝缘螺栓，两种固定螺栓均需有防松措施，固定可靠，无松动和窜动现象。 （10）绝缘夹件与引线固定处理措施，应垫以附加绝缘以防卡伤引线绝缘，引线固定用绝缘夹件的间距应考虑在电动力的作用下不致发生引线短路	（1）最新的《电力变压器检修导则》已删除引线与各部位之间的绝缘距离附表，但在"五通"里仍要求引线与各部位之间的绝缘距离应符合要求。 （2）包扎绝缘用皱纹纸宜用斜纹，不用竖纹纸。 （3）绕组抽头是分接开关连接处的整组引线应整齐，沿线支架应支撑有效，不得出现个别位置弧垂明显的严重情况
7	检查油箱（钟罩）	（1）检查油箱外部经过清洁，无明显渗漏点，无锈蚀，漆膜完整。 （2）法兰结合面无漆膜，光滑平整无锈蚀，并用酒精清洁大盖法兰结合面。 （3）油箱内侧磁屏蔽装置固定牢固，无放电痕迹，接地可靠，每组磁屏蔽装置编号完整。 （4）底座的器身定位钉（定位装置）不会造成铁心多点接地。 （5）更换所有密封件，胶垫压缩量为厚度的1/3，胶棒为1/2。 （6）大盖密封胶垫接头的搭接长度不小于胶垫厚度（宽度）的2倍	（1）新投运前、长途运输后、运输中未安装三维冲击记录仪、充气或充油保存的变压器未安装压力监视或压力低于0.01 MPa的应对油箱做密封试验。 （2）磁屏蔽装置在未发现连接问题或需要更换时，不宜加强紧固
8	工作结束	钟罩回吊前，收拾现场工具，冲洗场地出场地前清理鞋底油污，所有废弃杂物装入专门的垃圾袋	（1）注意核查铁心接地已恢复。 （2）夹件、绕组、铁轭上无遗漏工器具和零件

该项工作中绝大多数对器身的质量检查在电力变压器检修导则中要求的方式都是"目测",但由于不同厂家制造工艺的区别、新旧设备的差异、隐蔽位置、班组历年检修经验积累等原因,造成检查对象、位置,常见缺陷等问题的判断和处理方式不尽相同,出现吊罩后短短 1 小时即检查完成回装的情况。下文介绍一些器身检查工作容易忽略和检查的要点。

(1) 器身先清洗后检查,见图 2-40。

图 2-40 冲洗变压器器身及底座

(2) 引线防风摆处理,见图 2-41。凡在器身上工作要做好防外来脏污的措施。见图 2-42。

图 2-41 套管引线防风摆

图 2-42 在油箱内处理套管引线

(3) 检查油箱底座中异物见图 2-43;清理冲洗下来的脏污见图 2-44。

项目二 变压器检修 105

图 2-43 以往掉落在底座里的吊带

图 2-44 从变压器中清理出的金属碎屑

（4）低压侧引流线检查，见图 2-45。

图 2-45 低压端引流线（片）无明显放电、变色、缺损现象

（5）分接开关的选择器抽头连接处已做好连接标记，见图 2-46。

图 2-46 分接开关抽头连接已拆除并做标记

（6）铁心及铁轭完好情况，见图 2-47。
（7）夹件及螺杆防锈漆不能脱落，见图 2-48。

图 2-47　铁心紧密完好无折　　　　　图 2-48　螺杆防锈漆脱

(8) 夹件与铁心的绝缘板应紧实，无松动，无脱落，见图 2-49 和图 2-50。

图 2-49　夹件绝缘脱落　　　　　图 2-50　夹件与铁心间的绝缘板未松动

(9) 绕组外观及垫块应整齐，油道清晰可见，无脏污堵塞，见图 2-51 和图 2-52。

图 2-51　绝缘板伸出过长　　　　　图 2-52　垫块排列整齐

(10) 铁心、夹件、绕组汇接的隐蔽部位绝缘及垫块完好，见图 2-53。

项目二　变压器检修　　107

图 2-53　隐蔽部分垫块

（11）所有螺栓连接点检查有无放电痕迹，见图 2-54。

（12）不能出现绑扎脱落导致的引线坍塌，见图 2-55。

图 2-54　六角螺母角端对夹件放电　　图 2-55　高压侧引线坍塌

（13）油箱内壁磁屏蔽块装设牢固，无松动，应有数量编号，若有单独接地点则需测量屏蔽块对地绝缘，见图 2-56。

图 2-56　油箱磁屏蔽完好

（14）检查引线及支架走向平整、完好，见图2-57。

图 2-57　支架及引线走向平整

（15）油箱外壳检查并清洁脏污，特别注意检查隐蔽点，见图2-58。

图 2-58　清除油箱外壳隐蔽处的鸟窝

Task 1　Transformer Body Inspection after Core Hoisting

1.1　Work Tasks

Knowledge of what is involved in the transformer body inspection of a 110 kV main transformer after hoisting the bell hood and being able to apply the knowledge of the transformer body inspection of transformers of voltage classes 35 kV and below.

1.2 References

(1) *Electric Power Safety Working Regulations (Power Transformation) of State Grid Corporation of China* (Q/GDW1799.1-2013).

(2) Specialized teaching materials for the training of vocational competence of production skill personnel of the state grid of China: *Transformer Maintenance*.

(3) *Substation Maintenance Management Regulations of State Grid Corporation of China (Trial)—Volume 1: Detailed Rules for Maintenance of Oil Immersed Transformers (Reactors)*.

(4) *Regulations of State Grid Corporation of China on Management of Substation Operation and Maintenance (Trial)—Volume 1: Detailed Rules for Operation and Maintenance of Oil Immersed Transformers (Reactors)*.

(5) *Maintenance Guide for Power Transformer* (DL/T 573-2010).

(6) *Guide for the Operation and Maintenance of Tap Changers in the Power Transformer* (DL/T 574-2010).

1.3 On-site Climate and Work Requirements

(1) The environmental requirements are the same as the conditions for major repair of the transformer bell hood hoisting.

(2) The winding tap connections of the on-load tap changer selector have been removed and marked.

(3) The core of the on-load tap changer is hoisted out.

(4) The bell hood of the transformer has been hoisted out and placed on a number of load-bearing sleepers without the edges of the cap coming into direct contact with the ground.

1.4 Preparation for Work

1. Basic condition of equipment

The work is an individual work after the main transformer bell hood hoisting, and the transformer body (core, windings, leads, insulating support, and oil tank) components are static. According to the *Maintenance Guide for Power Transformer*, "visual inspection" is the main inspection item. Normal work does not change the content of its operation, structure, or wiring, mainly on the state of the designated object for inspection, cleaning, and testing.

2. Hazards and preventive and control measures

(1) Hazard: electric shock injury. Preventive and control measures are as follows.

① Insulating gloves are required when connecting the test leads of the megohmmeter.

② Insulating gloves should be checked.

③ Remember to discharge the battery after the insulation test.

(2) Hazard: slipping and falling. Preventive and control measures are as follows.

① Staff should properly wear non-slip shoes and shoe covers.

② Once the staff on the transformer body return to the ground, they need to replace the shoe covers with new ones before going up the ladder again.

③ Walking on the clamps of the transformer is allowed, but no running or jumping.

(3) Hazard: equipment falling. Preventive and control measures are as follows.

① When working on the transformer clamps, all personal appliances and parts are put in the tool bag.

② When sucking or flushing oil with the oil pump, pay attention to checking whether there are parts missing from the transformer base.

(4) Hazard: personal injury. Preventive and control measures are as follows.

When cleaning metal debris attached to the transformer body and transformer base, do not hold or grasp it directly with your hands. Please wear new cotton or cloth gloves to avoid scratches.

(5) Hazard: fire. Preventive and control measures are as follows.

① Ensure that all equipment in the fire fighting room is complete and has regular inspection records.

② No smoking on the work site. All fire-related work should be far away from oil hoses, oil purifiers, and other oil-related areas, and specialized areas for fire-related work should be available and marked.

3. Selection of work tools and materials

Work tools and materials for transformer body inspection and maintenance of 110 kV main transformer after core hoisting are shown in Table 2-3.

Table 2-3 Work Tools and Materials for Transformer Body Inspection and Maintenance of 110 kV Main Transformer after Core Hoisting

Category	Name	Specification and model	Quantity	Remarks
Specialized tools	Safety helmet		Provided by headcount	
	Safety belt		2	Used for flushing the core body, and transferring a small amount of transformer oil
	Oil hose and hoop	250—500 mm pipe diameter		In conjunction with oil pumps and purifiers

Continued

Category	Name	Specification and model	Quantity	Remarks
Specialized tools	Electronic megohmmeter			
	Oil tank			Oil storage
	Oil pump	250—500 mm pipe diameter	1 set	Depending on the pipe diameter of the oil hose
	Overhead light (or flashlight)			Can be worn on a safety helmet or carried on the back
	Ladders	3 m	1—2	For climbing on clamps
	Grounding wire			Electronic megohmmeter for external grounding
Personal tool	Slotted screwdriver			
	Cross-head screwdriver			
	Open spanner	8-10,10-12,17-19		Connector
	Adjustable spanner	One each of 150, 300, 600		
	White silk cloth			For wiping oil stains on the interior of the core body and oil chamber
	Cotton yarn			For wiping oil stains on floors
	Abrasive paper	#0 or #1	3 each	Corroded portion
Consumables	Transformer oil	#25		For flushing the core body
	Cotton gloves			
	Cloth gloves			
	Raincoat	Disposable, with hood		
	Shoe cover	Disposable		
	Alcohol	Industrial alcohol		For cleaning the bonding surface of the cap and flange

4. Division of labor among operators

The staff division of labor for the transformer body inspection and maintenance is shown in Table 2-4.

Table 2-4 Staff Division of Labor for Transformer Body Inspection and Maintenance of 110 kV Main Transformer after Core Hoisting

S/N	Job	Quantity	Job responsibilities
1	Person in charge of the work (guardian)	1	He/she is responsible for the division of labor for the work, site inspection before the work, the development of the work plan, the preparation of work tickets, the handling of work permits, the convening of pre- and post-shift meetings, the safety supervision during the work, the handling of emergencies during the work, the supervision of the quality of the work, and the summarization of the work
2	Operator	1	He/she is responsible for maintenance work, result verification, report, and record filling
3	Auxiliary operator	1	He/she assists with work requiring cooperation, written and photographic documentation of operations, and cleaning of the site after completion of the work

1.5 Operation Procedures

The overall inspection and maintenance operation procedures for the transformer body are shown in Table 2-5.

Table 2-5 Operation Procedures of Transformer Body Inspection and Maintenance of 110 kV Main Transformer after Core Hoisting

S/N	Scope of work	Operational steps and standards	Safety and technical considerations
1	Preliminary work	(1) The winding tap connections of the on-load tap changer selector have been removed and marked. (2) The bell hood of the transformer has been hoisted out and placed on a number of load-bearing sleepers without the edges of the cap coming into direct contact with the ground. (3) The oil hose for flushing is not rolled and is as straight as possible. (4) The oil drain hole in the base is opened to allow oil to be discharged during flushing	(1) Properly wear a safety helmet, work clothes, and labor protection gloves. (2) It is strictly prohibited for unrelated personnel and vehicles to enter the site
2	Inspection of work tools and consumables	(1) Prepare and neatly arrange the work tools according to the requirements, and check the appearance, performance and test certificate of the work tools, without omission.	(1) The appearance inspection of the work tools is qualified, with no damage, deformation, or malfunction, and the certificate of conformity is within the validity period.

Continued

S/N	Scope of work	Operational steps and standards	Safety and technical considerations
2	Inspection of work tools and consumables	(2) The electronic megohmmeter is checked for an on-off state, and the display, alarm, and discharge functions are all intact. (3) The disposable raincoat with hood is placed separately from the shoe cover and is not mixed with other dirty items. (4) The non-slip rubber gaskets on the bottom feet of the ladder are intact	(2) Disposable raincoats and shoe covers should be protected from pollution so that they do not cause secondary soiling of the transformer body stroke when worn and worked on the transformer body
3	Transformer body flushing	(1) An operator should properly wear a raincoat with cuffs and pant legs tied with rubber bands or white yarn bags. (2) Shoe covers should be worn while standing on as clean a tarp as possible before ascending the transformer body with the ladder. (3) The oil hose for flushing is handed over to the staff. (4) The oil pump supplies oil to the oil hose. If the power can be adjusted, wait until the oil has come out of the end of the tube held by the operator, and then gradually adjust the power. (5) The operator should flush the transformer core, windings, and clamps back and forth along the transformer body clamps. (6) Clean the upper clamps and accessible parts with a white silk cloth. (7) Once the operator has returned to the ground, other staff members may then flush the sides of the transformer body	(1) When going up and down the ladder, it is necessary to be supported by a specially-assigned person because the ladder is resting on the winding. (2) Oil for flushing can be used from the transformer oil discharged. (3) Oil pumps with an adjustable power function should avoid one-time high-power oil discharge. Sudden straightening of the oil hose should be avoided to cause the operator to drop it unsteadily or knock the operator down. (4) Since a large amount of dirt and metal debris will be brushed off after flushing, the flushed oil should be stored in a separate tank or container and should not be filtered for use. (5) When the operator flushes above the transformer body, particular attention should be paid to flushing the oil ducts, clamp gaps, upper yoke, and winding cores. (6) If the operator has returned to the ground and is to go up again, new shoe covers are required

Continued

S/N	Scope of work	Operational steps and standards	Safety and technical considerations
4	Core inspection	(1) The appearance of the core should be checked for completeness, with primary attention paid to the uniformity of the edge layers and the tightness of the lamination. Its surface and edges should not be warped or bent. Clean the surface of the core with a clean white cloth to remove the oil contamination, iron dust, and other dirt, especially not a wide range of gaps between the layers of the core, or bending short-circuits at the end of the lapped portion of the core. (2) The oil ducts should be observed to be neatly aligned and free of instability such as dislodged pads, blockage, or sliding. (3) The insulation between the core and the upper and lower clamps and pressboards should be intact, and there should be no rupture caused by pressure piling or abrasion. (4) The antirust paint on the core insulation, clamps, and other metal parts should be free from discharge burns and traces of overheating. (5) There should be a point of reliable grounding between the core groups, clamps, through bolts, and steel drawstring (or epoxy resin tapes). (6) The core grounding piece should be inserted firmly, and its exposed part should be insulated and wrapped to prevent the core from short-circuiting. (7) The insulation of the core and clamps to the ground should be measured with a megohmmeter of gear 2,500 V or higher	(1) It is prohibited to knock down curled cores or uneven parts as required by the old version Guide, as this may damage the core surface insulation. (2) Special attention should be paid to the tip discharge marks of each metal part. In addition, the chamfered portion of the edge of the nut may also discharge the clamp. (3) The *Maintenance Guide for Power Transformer* requires that the core and clamp insulation against the ground is greater than 1MΩ, but the "Five General Regulations" require it to be ≥ 100 MΩ (newly commissioned 1,000 MΩ) (750 kV and below), which needs to be implemented in accordance with the latter

Continued

S/N	Scope of work	Operational steps and standards	Safety and technical considerations
5	Winding inspection	(1) (If available) The interphase partitions and enclosures should be checked for damage, discoloration, deformation, or traces of discharge. (2) (No enclosure) The windings should be visually inspected to ensure that they are clean, and free of oil dirt, deformation, overheating, discoloration, or traces of discharge. (3) The winding as a whole should be free of tilting, displacement, and no obvious popping out of wires. (4) The flushed oil duct should be free of oil dirt and other impurities accumulated. (5) The upper and lower pads of the windings and the interstage strips should not be displaced or loosened after flushing, and should be neatly arranged. The length of the winding exposed outside the pads should not exceed the thickness of the winding wires. (6) The insulation status of the winding should be checked by finger pressing	In addition to frontal observation of the winding as a whole, it should also be observed on the side of the clamp without overall left and right, up and down misalignment, otherwise, it will cause the bottom of the winding to overhang, the winding collapse, or the core and winding short-circuiting
6	Lead and insulating support inspection	(1) The leads should be checked visually for no broken strand damage or traces of overheating. (2) Visually or by hand pinching, check whether there is any stress concentration in the tensile and bending stress concentration areas such as the position of the winding's wire leads, the position of the insulating clamps, the welded parts, and the connectors of the tap changer leads. (3) The wrapping insulation of the leads shall be visually inspected for completeness, without deformation, wrinkling, brittleness, breakage, broken strands, or discoloration.	(1) The most recent *Maintenance Guide for Power Transformer* has deleted the schedule of insulation distance between the leads and the parts, but the "Five General Regulations" still require that "the insulation distance between the leads and the parts shall comply with the requirements". (2) The insulation shall be wrapped with crepe paper, preferably with a diagonal pattern, not with a vertical pattern.

Continued

S/N	Scope of work	Operational steps and standards	Safety and technical considerations
6	Lead and insulating support inspection	(4) The lead joints and joint surfaces should be checked to see if they are flat and smooth, without burrs, or overheating discoloration. The area of joints should be more than 3 times its lead cross-section. (5) The cable-through bushing shall be checked for cable-through leads, which shall be wrapped in one layer by a half stack of white galloon. (6) The high-current leads shall be checked on the LV side to see if they are wrapped and insulated, and if they are not wrapped, they shall be wrapped in accordance with the regulations. (7) The insulation distance between the leads and the components shall be checked. (8) The insulating support shall be checked for looseness, damage, or displacement, and shall be free from breakage, cracks, bending deformation, or burns. (9) The fixing of the insulating support to the iron clamps should be checked. Steel bolts are available and insulated bolts should be used for fixing the insulating parts to the insulating support. Both fixing bolts should have anti-loosening measures, reliably fixed, with no loosening, or movement. (10) For fixing measures for insulating clamps and leads, additional insulation should be provided to prevent jamming the lead insulation. The spacing of the insulating clamps used for fixing the leads should be taken into account so that no short-circuiting of the leads occurs under the action of electric power	(3) The whole set of leads from the winding tap to the on-load tap changer connection shall be neat. Supports along the leads shall be supported effectively and there shall be no obvious severe sag at individual locations

Continued

S/N	Scope of work	Operational steps and standards	Safety and technical considerations
7	Oil tank (bell hood) inspection	(1) The exterior of the tank should be checked to see if it has been cleaned, with no obvious leakage points, free of rust and corrosion, and has a complete paint film. (2) There is no paint film on the flange bonding surface, smooth and flat, with no corrosion, and alcohol is used to clean the bonding surface of the cap and flange. (3) The magnetic shielding device on the inside of the tank should be firmly fixed, without discharge marks, and grounded reliably. Each group of magnetic shielding devices should be numbered completely. (4) The transformer body locating pins (locating devices) of the base will not cause multiple points of grounding of the core. (5) All seals should be replaced. The compression of the rubber gasket is 1/3 of its thickness and 1/2 for the rubber rod. (6) The lap length of the cap sealing gasket joint is not less than 2 times the thickness (width) of the gasket	(1) If the 3D impact recorder is not installed before new commissioning, after long-distance transportation, or during transportation, and if the pressure monitor is not installed in the transformer that is inflated or filled with oil for preservation or if the pressure is lower than 0.01 MPa, the sealing test should be done on the oil tank. (2) Magnetic shields should not be tightened more than necessary without identifying connection problems or the need for replacement
8	End of work	Prior to bell hood back hoisting, it shall pack up tools on the site, rinse the on-site oil contamination, and clean the oil dirty from the soles of shoes before leaving the site. All waste debris should be placed in special trash bags	(1) Care should be taken to verify that core grounding has been restored. (2) There shall be no missing work tools or parts on the clamps, windings, or yoke

In this work, most of the quality inspection of the transformer body is "visual inspection" in the maintenance guide of power transformer. However, due to the difference of manufacturing process of different manufacturers, the difference of new and old equipment, the concealed location, the accumulated maintenance experience of the team over the years and other reasons, the judgment and treatment of the inspection object, location, common defects and other problems are not the same. Check the completion of reassembly within 1 hour after the occurrence of lifting cover. The following introduces some key points that are easy to be ignored and checked in machine body inspection.

(1) Clean the body before inspection. See Fig. 2-40.

Fig. 2-40　Flush the Transformer Body and Base

(2) The lead should be wind proof. See Fig. 2-41. Anyone who works on the body should take measures to prevent foreign pollution. See Fig. 2-42.

Fig. 2-41　Wind Proof Swing of Casing Lead　　Fig. 2-42　Dealing with Casing Lead in Oil Tank

(3) Check the foreign matters in the oil tank base, see Fig. 2-43. And clean up the dirty stains, see Fig. 2-44.

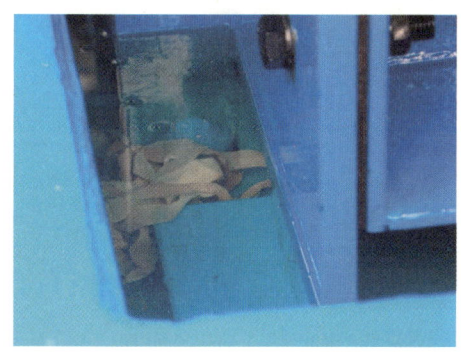 

Fig. 2-43　The Sling That Used to Fall in the Base　　Fig. 2-44　Metal Scraps Cleaned from Transformers

(4) Low pressure side guide line inspection. See Fig. 2-45.

Fig. 2-45 Low Pressure End Drainage Line (Film) has No Obvious Discharge, Discoloration and Defect

(5) The selector tap connection of the tap changer has been marked. See Fig. 2-46.

Fig. 2-46 Tap Connection of Tap Changer Removed and Marked

(6) The integrity of the core and yoke is shown in Fig. 2-47.

Fig. 2-47 The Core Is Tight and Unfolded

(7) The clamp and screw anti rust paint can not fall off. See Fig. 2-48.

Fig. 2-48　Screw Anti Rust Paint Falling Off

(8) The insulation board between clamp and core should be tight without looseness or falling off. See Fig. 2-49 and Fig. 2-50.

Fig. 2-49　Clamp Insulation Falling Off

Fig. 2-50　The Insulating Plate Between Clamp and Core Is not Loose

(9) The winding appearance and cushion block should be neat. The oil passage should be clear and visible, and there should be no dirt and blockage. See Fig. 2-51 and Fig. 2-52.

Fig. 2-51   Insulation Board Extended Too Long   Fig. 2-52   The Cushion Blocks Are Arranged in Order

(10) The hidden parts of the iron core, clamps and winding are insulated and the blocks are intact. See Fig. 2-53.

Fig. 2-53   Concealed Part of Pad

(11) Check all bolt connection points for discharge traces. See Fig. 2-54.
(12) Lead collapse caused by binding falling off is not allowed. See Fig. 2-55.

Fig. 2-54　Hex Nut Corner End to Clamp Discharge　　Fig. 2-55　High Voltage Side Lead Collapse

(13) The magnetic shield block on the inner wall of the oil tank should be installed firmly without looseness and should have a number. If there is a separate grounding point, the insulation between the shield block and the ground should be measured. See Fig. 2-56.

Fig. 2-56　Magnetic Shield of Oil Tank Is Intact

(14) Check whether the lead wire and bracket are in good condition. See Fig. 2-57.

Fig. 2-57　The Direction of Support and Lead Is Flat

(15) Check and clean the oil tank shell, pay special attention to the concealed points. See Fig. 2-58.

Fig. 2-58　Remove the Bird's nest from the Cover of the Oil Tank

任务二　M 型有载分接开关机构箱检修

一、工作任务

（1）完成 M 型有载分接开关机构维护检查的项目和内容。

（2）完成 M 型有载分接开关机构挡位调试和功能验证。

（3）熟悉有载分接开关二次回路图的控制功能识读。

二、引用标准

（1）《国家电网公司电力安全工作规程（变电部分）》（Q/GDW1799.1—2013）。

（2）国家电网公司生产技能人员职业能力培训专用教材——《变压器检修》。

（3）《国家电网公司变电检修管理规定（试行）第 1 分册油浸式变压器（电抗器）检修细则》。

（4）《国家电网公司变电运维管理规定（试行）第 1 分册油浸式变压器（电抗器）运维细则》。

（5）《电力变压器检修导则》（DL/T 573—2010）。

（6）《变压器分接开关运行维修导则》（DL/T 574—2010）。

（7）《分接开关第 1 部分：性能要求和试验方法》（GB 10230.1—2007）。

（8）《分接开关第 2 部分：应用导则》（GB/T 10230.2—2007）。

三、工作要求

（1）按照《有载分接开关运行维修导则》要求机构箱每年应进行一次检修、检查工作。
（2）一般机构的使用在海拔 2000 m 以下，如超过或特别订制型式需做特殊说明。
（3）环境温度不高于 + 55 ℃，不低于 - 25 ℃。
（4）机构与地平面的垂直度偏差不超过 5%。
（5）机构运行场地无严重尘埃及其他爆炸性、腐蚀性气体。

四、工作准备

1. 设备基本状况查勘

（1）变压器本体及分接开关附近无渗漏油缺陷，若有则要先处理渗漏油。
（2）记录工作前的分接开关机构显示挡位、远控显示挡位、分接开关芯体窗口显示的实际挡位，三者应该一致，不一致则需要在工作中判断缺陷出现位置及处理。
（3）记录分接开关机构当前挡位下的计数器数值。
（4）机构传动联杆目测无错位，无零件脱落。
（5）机构传动联杆目测无明显变形，无大面积锈蚀，若有则需要更换。
（6）核对机构铭牌，特别注意机构型号和厂家。

2. 危险点及预控措施

（1）危险点——触电伤害。预控措施如下：
① 主变未停电时的机构维护检查不能爬上主变观察挡位，只能从远控端观察挡位。
② 确定临时电源位置，电缆放线尽量从工作区域以外绕行。
③ 操作前用万用表验证机构电源侧已断开。
④ 班组若自备开关箱注意检查接线和接地，确认空开处于断开位置，通电时，先合上通临时电源空开，再合上开关箱空开，最后合上机构箱空开。
⑤ 所有工作用电缆接线需检查绝缘完好状况，若有破裂，用黄色绝缘防水胶带多层包扎，并在下面垫干燥木板。
（2）危险点——防止滑倒摔伤。预控措施如下：
① 工作人员应正确穿戴防滑鞋。
② 地面若有油污，进出工作现场、上下梯子需要清洁鞋底。
③ 为防止机构零件掉落入池内无法寻找，在机构箱下的防火池地面上根据操作范围适量铺设白布或蛇纹布，尽量不用油布，零件掉落在油布上可能被弹开掉入防火池，且在沾油的油布上行走更容易滑倒。
④ 在防火池内架设梯子或绝缘高凳，必须有专人扶持。

3. 工器具及材料选择

M 型有载分接开关机构箱检修工器具及材料见表 2-6。

表 2-6 M 型有载分接开关机构箱检修工器具及耗材

类别	名称	规格型号	数量	备注
专用工具	安全帽		3 顶	
	绝缘手套		1 副	
	兆欧表（或电子式）	500 V 或 1000 V	1 套	
	万用表		1 套	
	配电箱	便携式	1 套	
个人工具	平口改刀		1 把	回路端子
	十字改刀		1 把	拆面板
	呆扳手	8-10,10-12,17-19	各 2 把	拆除连接件
	活络扳手	300 mm	1 把	辅助活络扳手使用
	工具盘		1 张	
耗材	砂纸	#0 或#1	各 3 张	锈蚀部分
	润滑脂	锂基脂	1 包	机构外露齿轮，不能使用钙基脂
	鬃毛刷		1 把	
	白绸布		若干 m²	
	白布		若干 m²	

4. 作业人员分工

M 型有载分接开关机构箱检修人员分工见表 2-7。

表 2-7 M 型有载分接开关机构箱检修人员分工

序号	工作岗位	数量	工作职责
1	工作负责（监护人）	1	负责本次工作的人员分工、工作前的现场查看、作业方案制定，工作票的编制，办理工作许可手续，召开工作班前班后会，负责作业工程中的安全监督、工作中突发情况处理，工作质量监督，工作后的总结
2	操作人员	1	负责检修工作，结果核查，报表、记录填写
3	辅助作业人员	1	协助需要配合的工作，操作过程中的书面和照相记录，工作完成后现场清洁

五、工作程序

M 型有载分接开关机构箱检修流程见表 2-8。

表 2-8　M 型有载分接开关机构箱检修流程

序号	作业内容	作业步骤及标准	技术及安全质量要求
1	前期准备工作	（1）规范填写和签发工作票，正确履行工作票手续。 （2）现场查勘必须由 2 人进行。现场核对分接开关对应变压器名称、分接开关类型、当前运行挡位。 （3）正确装设安全围栏，悬挂标示牌	（1）正确穿戴安全帽、工作服、劳保手套。 （2）不得在分接开关联杆未断开或主变压器间隔未停电情况下进行挡位变换。 （3）操作分接开关前记录机构和远控显示器当前运行挡位。 （4）严禁无关人员、车辆进入作业现场
2	检查工器具	（1）将工器具按要求准备齐全并摆放整齐，检查工器具外观、性能和试验合格证，无遗漏。 （2）对绝缘手套做充气试验，方法正确。 （3）对兆欧表做短路试验和开路试验，并调零（机械式）。 （4）对万用表做通断检查，显示和蜂鸣报警均完好	（1）工器具外观检查合格，无损伤、变形、失灵现象，合格证在有效期内。 （2）绝缘手套、兆欧表、万用表试验合格
3	记录数据	（1）记录机构铭牌参数。 （2）记录机构显示当前挡位 A。 （3）记录远控（液晶或电子显示）挡位 B。 （4）记录机构当前计数器数字 N。 （5）A 和 B 应一致	（1）原则上记录过程不允许打开机构箱。 （2）若机构观察窗口老化严重观察不清，可开箱记录，此时禁止碰触内部任意元件
4	停电（情况1）	（1）确认被检修主变压器间隔已停电，安措已完备，许可开始工作。 （2）若仅对主变压器停电，则确认高低压侧引线均已断开，接地线已设好，确认主变压器外壳、中性点、铁心接地完好，并验电确认	（1）不得约时停电。 （2）再次确认停电间隔所有设备名称和编号准确无误
5	不停电（情况2）	（1）对分接开关联杆做验电工作，不得带电。 （2）戴绝缘手套断开与机构相连的传动联杆，并完全取下	（1）严禁爬上变压器工作。 （2）变压器爬梯上"禁止攀登"标示牌应明显。 （3）变压器爬梯锁须锁好

续表

序号	作业内容	作业步骤及标准	技术及安全质量要求
6	机构电源检查	（1）确认机构二次电源在集控箱中的空开已断开，并验电确认。 （2）若二次电源线已取下，则确认自带配电箱或现场临时电源接线完好，空开保护功能完好	（1）在集控箱中的对应空开上悬挂"有人工作，禁止合闸"小标示牌。 （2）临时电源线绝缘表皮无破裂，若有破裂则用黄色绝缘胶带往返包扎至少两层，布置于人员流动较少的区域，避免踩踏。 （3）必须得到呼唱指令才能进行机构送电操作
7	确认二次回路电源断开	（1）确认箱内二次回路空开已断开。 （2）将控制旋钮从"远方"调至"停止"。 （3）将端子排熔断器断开，并取下检查熔断器通断	（1）在进行电动调试前禁止将控制旋钮调至"就地"位置。 （2）熔断器若碰撞或掉落后熔丝可能断裂，则取下熔断器检查后应装回原位，若未装回，则用纸或布包扎后放置于不与其他硬质物品直接碰撞的地方
8	取下遮蔽面板	（1）开箱后取下遮蔽面板。 （2）检查操作手柄在面板或机构门内侧正常悬挂	面板上有各种标记，须在取下后放置于不与刚性物品碰撞的地方，避免掉漆和划伤标记
9	检查机构密封	（1）检查机构门密封胶条在卡槽内，无脱落，按压应具有良好弹性。 （2）检查机构底部电源和控制电缆进出孔封堵胶密封完好，未完全硬化变脆	未确认机构电源断开前禁止用手检查进出孔封堵胶
10	清洁	（1）用毛刷和白绸布配合，清洁机构内部和各元件上的灰尘。 （2）清洁后确认无刷毛卡落在元件缝隙中	（1）毛刷应软硬适中，避免过硬的刷毛卡在元件的螺丝压接面或其他缝隙中。 （2）毛刷不能已沾油或沾脂
11	检查齿轮盒	（1）用操作手柄升降各一个挡位，感觉无不均匀卡涩，无连续或间断碰撞声。 （2）用白绸布擦拭齿轮盒底部密封部位，无水和油渗出。 （3）齿轮盒上应有出厂时的"禁止注油"标示。 （4）两人配合：一人匀速转动手柄，一人用毛刷沾润滑脂均匀涂抹于齿轮盒上方的外露齿轮上	（1）齿轮盒沿下方出厂时为焊接密封，严禁工作人员自行拆开检查。 （2）禁止向齿轮盒内自行注入润滑油

续表

序号	作业内容	作业步骤及标准	技术及安全质量要求
12	检查电气连接	（1）目测和手动配合检查机构内所有电气元件接线良好，无松动，接线端头无明显放电痕迹。 （2）检查所有端子编号完好清晰，对缺损和不清晰的照相并记录。 （2）检查所有行程开关的动作情况和触点连接声音完好。 （3）检查所有接触器触点弹跳正常	检查期间禁止向机构通电
13	测量加热电阻	用万用表测量加热电阻阻值，符合电阻上标记的数值，如 $1000×（1±2\%）Ω$	测量时断开加热电阻外部接线，测量完后需立即恢复
14	测量回路绝缘电阻	（1）断开机构外部电源电缆。 （2）合上端子排熔断器。 （3）确认机构门内侧的接地连接完好。 （4）用 500 V 或 1000 V 兆欧表进行回路绝缘电阻测量，绝缘电阻不小于 $1 MΩ$。 （5）根据需要可断开熔断器，分别测量带电机的主回路绝缘电阻和控制回路绝缘电阻	（1）测量期间禁止向机构通电。 （2）配合测量人员需戴绝缘手套，在放电完成之后才能取下。 （3）在电机未出现缺陷时，不建议开盖对电机进行绝缘测试
15	挡位转动过程检查	任意位置手动升降挡位一个循环，能从手柄的卡涩程度和声音上体验到"选择器脱扣—选择器合扣—切换开关完成切换—电机行程开关弹跳断电"的四个过程，即为转动过程机械动作正常。（具体方法见后续相关理论知识）	检查期间禁止向机构通电
16	挡位调零	手动操作挡位至整定档，检查升降两档圈数之差为零，否则应进行联结校验。（具体方法见后续相关理论知识）	检查期间禁止向机构通电
17	检查机械极限限位	（1）手动操作挡位分别至最高挡和1挡，继续摇动直至到达机械限位位置，无法转动为止。 （2）检查极限位螺丝闭锁连接完好	检查期间禁止向机构通电
18	机构通电	（1）确认当前挡位处于切换完成状态，即红线指示在窗口中间。 （2）确认回路断电的行程开关凸轮已弹开。 （3）确认空开处于断开状态。	（1）确认工作完成前禁止通电。 （2）通电必须呼唱并得到回复。 （3）空开断开状态指示灯未亮则用万用表核实，确认失电后再进行检查

续表

序号	作业内容	作业步骤及标准	技术及安全质量要求
18	机构通电	（4）发出通电的呼唱指令，并得到通电的操作回复。 （5）确认空开指示灯亮起。 （6）合上空开，并将控制旋钮从"停止"调至"就地"	
19	检查电气极限限位	（1）手动至最高挡位和最低挡位，自动停止后继续按电动升或降按钮，最高挡位不会继续升，最低挡位不会继续降。 （2）在升降过程中，未按停止按钮时，不会发生跳空开断电的情况，否则应消缺	机械限位检查合格前禁止直接通电进行电气限位检查，否则若挡位有偏差，可能导致与芯体连接的联杆接头被打断
20	检查升降逻辑和过程	（1）在极限限位完好的情况下，电动分接变换一个循环。 （2）循环过程中检查启动按钮应准确可靠，升降挡逻辑正确。 （3）循环过程中任意位置检查紧急停车按钮应可靠动作。 （4）对具有超越节点功能的机构检查超越节点可靠工作，即具有9a、9b、9c挡位的开关，电动循环时在9a、9c不会停止，在其他挡位正常停止。 （5）检查在每个挡位自动停止时，窗口中的红线均在靠近中间位置	（1）每次电动操作间隔5 s，待接触器线圈放电后再继续升降。 （2）电动期间若使用了操作手柄，在下次电动时确认手柄已取下，避免短路情况下手柄被甩出伤人
21	检查计数器计数	根据数据记录，此时计数器增加计数应与本工作中挡位切换次数吻合	计数器计数错误或失效，会影响根据切换次数来确定的机构小修、开关大修的周期
22	检查时间继电器功能	（1）在电动循环能在每一挡正确自动停止的前提下，将时间继电器时间调至5 s以下，再次电动，机构能在完成一挡转动前就跳空开停止。 （2）恢复时间继电器时间，具有超越节点功能的机构一般时间设置为12.5 s，无越节点功能的机构一般设置时间为7.5 s	（1）无超越节点功能的机构，时间继电器时间设置原则应满足大于一个挡位电动切换时间，小于两个挡位电动切换时间。 （2）有超越节点功能的机构，时间继电器时间设置原则应满足大于超越节点时至少两个挡位电动切换时间，小于三个挡位电动切换时间

续表

序号	作业内容	作业步骤及标准	技术及安全质量要求
23	检查手动电动互锁	在通电并空开合上情况下，插入操作手柄，空开立刻跳闸，并且在手柄插入期间，空开无法合上	该功能一旦失效，可能操作时误碰电动，导致操作手柄被直接甩出伤人、伤物
24	远控功能检查	（1）将控制旋钮从"就地"调至"远方"。 （2）从远控端呼唱并操作一个电动循环。 （3）检查在每个挡位自动停止时，窗口中的红线均在靠近中间位置。 （4）每个挡位自动停止时远控显示挡位与机构内窗口显示挡位一致	远控功能如失效或显示不全，对运维操作和后台监控有影响，若机构和开关本体功能完好，就地对机构进行操作及开关切换过程不受影响
25	恢复挡位	（1）根据数据记录，电动或手动恢复至初始挡位。 （2）确认远控显示挡位与机构内窗口显示挡位一致	完整确定应该是远控显示挡位、机构内窗口显示挡位、开关顶部观察窗口显示挡位三者一致，但在未停电情况下不能观察开关顶部挡位，因此要求在"检查升降逻辑和过程"项目时切实做到一个循环操作，才可能发现是否有开关顶部挡位和机构挡位不一致
26	断开电源	断开检修电源或集控箱中的二次电源	
27	恢复联杆连接	（1）恢复开始拆除的传动联杆连接。 （2）确认联杆螺栓齐全，并上下杆部连接螺栓紧缩片已压紧	恢复联杆连接时若使用绝缘高凳或人字梯，应注意监护，禁止身体探入变压器上部未停电区域
28	恢复遮蔽面板	（1）再次确认机构内和机构顶部无遗漏零件和工具。 （2）恢复遮蔽面板。 （3）恢复操作手柄悬挂位置	（1）注意封堵胶和箱底内侧容易粘连和遗漏零件。 （2）操作手柄禁止直接横放在箱底空间

变压器本体附件外观状态评价扣分表见表 2-9。

表 2-9　变压器本体附件外观状态评价扣分表

评判小项	检查方式	评判原则
一、油箱外观		
设备出厂铭牌齐全、清晰可识别	现场检查及查阅资料（包括 110 kV 及以上本体"油温-油位"曲线）	无设备出厂铭牌的，扣 0.5 分； 铭牌锈蚀或不清晰程度超过 1/3 的，扣 0.2 分； 铭牌安装部位不合理，现场无法准确查看的，扣 0.2 分铭牌部位较高无法准确查看，但可提供照片的不扣分

续表

评判小项	检查方式	评判原则
运行编号标志清晰、正确可识别	现场检查（包括散热器编号）	无运行编号及设备双重编号不完整、无法正确识别的，扣 0.5 分； 运行编号标志不清晰程度超过 1/3 的，扣 0.2 分
相序标志清晰、正确可识别	现场检查	相序标志无法正确识别的，扣 0.5 分； 相序标志不清晰程度超过 1/3 的，扣 0.2 分
二、温度计		
温度计指示结果	现场检查及查阅资料（包括绕组和油面温度计）	现场温度计指示、控制室温度显示装置、监控系统显示的温度应基本保持一致，最大误差不超过 5K，绕组温度不应低于油温，不符合要求的，发现 1 次扣 1.5 分
相间温差		对于单相变压器，同组设备不同相别温度差应小于 10K，不符合要求的，扣 1.5 分
温度计引出线固定		温度计引出线固定良好，应设反水弯等防止雨水倒灌措施，绕线盘半径不小于 50 mm，不符合要求的，发现 1 次扣 1.5 分
防雨措施（户内除外）		加装防雨罩，本体及二次电缆进线 50 mm 应被遮蔽，45°向下雨水不能直淋，不符合要求的，扣 1.5 分
温度计表盘指示		温度计指示清晰，无进水现象，历史最高温度指示正确，表盘定值位置与定值单整定基本一致，不符合要求的，发现 1 次扣 1.5 分
温度记录		按照变电站全面巡视周期记录温度、对应负荷和环境温度，不符合要求的，扣 1.5 分
三、油枕		
油位指示应符合"油温-油位曲线"	现场检查	供应商有标准时，按标准执行，无标准时按油位应在标准曲线 −10%～10% 范围内，不符合要求的，扣 1.5 分
油位计应加装防雨罩（户外变压器），内部无油垢，油位清晰可见，可在运行中就地读取		油位计观察窗应能够清晰识别，油位计加装防雨罩，本体及二次电缆进线 50 mm 应被遮蔽，45°向下雨水不能直淋，不符合要求的，扣 1.5 分
四、吸湿器		
玻璃罩杯油封完好，能起到长期呼吸作用	现场检查	玻璃罩杯无破损，密封完好无进水，呼或吸状态下，内油面或外油面应高于呼吸管口，油杯内油位不应过高，不符合要求的，扣 1.5 分

续表

评判小项	检查方式	评判原则
使用变色吸湿剂,罐装至顶部1/6~1/5处,受潮吸湿剂不超过2/3	现场检查	未使用变色吸湿剂的,扣1.5分;吸湿剂剂量不符合规定的,扣1分。
吸湿剂上部不应被油浸润,无碎裂、粉化现象		不符合要求的,扣1.5分
连通管应整体清洁、无堵塞、无锈蚀,与油枕旁通阀门位置应正确		不符合要求的,扣1.5分
免维护吸湿器电源应完好,加热器工作正常启动定值小于RH 60%或按厂家规定		不符合要求的,扣1.5分
五、阀门		
阀门必须根据实际需要,处在关闭和开启位置	现场检查	不符合要求的,发现1次扣1.5分
指示开闭位置的标志清晰正确		不符合要求的,扣1.5分
六、噪声		
运行中应无异常噪声	现场检查	不符合要求的,发现1次扣1.5分
七、锈蚀		
本体及组件金属部位无明显锈蚀	现场检查	锈蚀面积不应大于400 mm^2,发现1次扣0.1分;螺栓锈蚀的,导电部位螺栓锈蚀,扣0.1分,非导电部位螺栓锈蚀,扣0.02分;所有部位的锈蚀,最多扣0.5分
八、渗漏		
本体及组件负压区无渗漏油	现场检查	负压区包括潜油泵进油口、有载开关在线滤油装置油筒至油泵进油口、储油柜顶部、套管储油柜等接近或高于油面的区域等,不符合要求的,发现一处扣2分
本体及组件正压区无渗漏油		不符合要求的,属于严重及以上缺陷的,发现一处扣1分;属于一般缺陷的,发现一处扣0.5分
九、本体端子箱		
控制线电缆芯无外露,如采用多芯软铜线,应压接绝缘接线柱	现场检查	不符合要求的,扣1.5分
端子箱箱体接地、箱内二次接地、箱门与箱体的接地连接线连接良好		未进行接地连接的,扣1.5分;采用接地线不规范或接地部位不可靠、锈蚀的,扣0.5分

评判小项	检查方式	评判原则
驱潮装置和加热升温装置工作正常，温湿度整定值设定按现场运行规程投退，若使用加热器其位置应与各元件、电缆及电线的距离应大于 50 mm	现场检查	不符合要求的，扣 1.5 分
驱潮装置和加热升温装置工作正常，温湿度整定值设定及驱潮装置和加热升温装置投退均按现场运行规程执行		不符合要求的，扣 1.5 分
箱门密封、内部封堵应良好，无进水、受潮、积灰、存在异物现象，如使用荧光灯管门灯，应加装防护罩		不符合要求的，扣 1.5 分
十、套管		
油位指示	现场检查	油位或气体压力正常，油位或压力计就地指示应清晰，便于观察（可借助长焦相机或望远镜），油套管垂直安装油位在 1/2 以上（非满油位），倾斜 15°安装应高于 2/3 至满油位，不符合要求的，发现 1 次扣 1.5 分
绝缘子无碰损或开裂，法兰无开裂，单个缺釉不大于 25 mm^2，釉面杂质总面不超过 100 mm^2		不符合要求的，发现 1 次扣 1.5 分
无放电、严重电晕和电腐蚀现象		•不符合要求的，发现 1 次扣 1.5 分
抱箍、线夹应无裂纹现象	现场检查	不符合要求的，发现 1 次扣 1.5 分
户外压接型设备线夹，朝上 30°~90°（ϕ400 及以上）安装时应配钻直径 6~8 mm 的排水孔，线夹不应采用铜铝对接过渡线夹，铜铝对接线夹应制定更换计划。110（66）kV 及以上电压等级变压器套管接线端子（抱箍线夹）应采用 T2 纯铜材质热挤压成型。禁止采用黄铜材质或铸造成型的抱箍线夹	现场检查及查阅资料	不符合要求的，发现 1 次扣 1.5 分

续表

评判小项	检查方式	评判原则
引线应无散股、扭曲、断股现象	现场检查	不符合要求的,发现1次扣1.5分
套管本体应无温度异常	现场检测	不符合要求的,发现1次扣1.5分
应无接头发热		不符合要求的,发现1次扣1.5分
实际爬电比距与污区等级相匹配	查阅资料	不符合要求的,发现1次扣1.5分
金属法兰与瓷件浇装部位黏合应牢固,防水胶完好,喷砂均匀,无明显电腐蚀	现场检查	不符合要求的,发现1次扣1.5分
十一、有载分接开关		
油位指示应清晰、准确,便于观察	现场检查	不符合要求的,扣1.5分
油位应正常,不应过高或过低		不符合要求的,扣1.5分
玻璃罩杯油封完好,能起到长期呼吸作用	现场检查	玻璃罩杯无破损,密封完好无进水,呼或吸状态下,内油面或外油面应高于呼吸管口,不符合要求的,扣1.5分
使用变色吸湿剂,罐装至顶部1/6~1/5处,受潮吸湿剂不超过2/3		未使用变色吸湿剂的,扣1.5分;吸湿剂剂量不符合规定的,扣1分
吸湿剂不应自上而下变色,上部不应被油浸润,无碎裂、粉化现象		不符合要求的,扣1.5分
免维护吸湿器电源应完好,加热器工作正常启动定值小于RH 60%或按厂家规定		不符合要求的,扣1.5分
交接、吊检后应进行切换特性试验	查阅资料	未进行切换特性试验的,发现1次扣3分
分接转换开关本体应配装机械限位装置,机构箱内应有机械限位和电气闭锁,出现滑挡、飞车缺陷应采取处理措施	现场检查及查阅资料	没有限位装置和电气闭锁的,扣2分;出现滑挡、飞车缺陷未采取处理措施的,发现一次扣1分
挡位分接指示应与控制室、本体、操作机构一致	现场检查	现场实际挡位指示与其他指示不一致的,扣1.5分
在线滤油装置滤芯应按根据油压和运行时间规定进行更换	现场检查及查阅资料	未按周期进行滤芯更换的,扣1.5分;无更换检修记录的,发现1次扣1分
在线滤油装置压力应正常,无渗漏油	现场检查及查阅资料	压力不正常的,扣1.5分;表计接头处渗漏油的,扣1分
在线滤油装置极寒条件下应配置防冻措施,应有自启动加热保温装置		"极寒条件"是指分接开关内油温低于0℃及以下的情况,不符合要求的,扣1.5分

项目二　变压器检修

续表

评判小项	检查方式	评判原则
应按根据评价结果开展试验	查阅资料	未按照《输变电设备状态检修试验规程》按期进行试验的,发现1次扣3分
油耐压应合格	查阅资料(最近1次试验报告)	试验数据不符合要求的,扣3分
十二、无载分接开关		
有挡位输出的无励磁分接开关分接挡位指示应与控制室一致	现场检查	现场实际挡位指示与其他指示不一致的,扣1.5分(如挡位无引出则不扣分)
十三、冷却装置		
风机或油泵应有轮换运行记录	查阅资料	冷却装置轮换运行记录不规范的,扣0.1分;轮换运行记录内容不齐全的,缺少一次扣0.5分,最多不超过2分
管状结构变压器冷却器每年应至少进行1次冲洗,并在大负荷前进行	查阅资料	没有按照规定进行冲洗且没有检修记录的,扣1.5分;检修记录不完整的,缺少一次扣0.5分。(完整性由2015年开始检查)
变电站监控后台及调度监控能够监视主变冷却装置运行情况,冷却器全停后能够正确上传信号	现场检查及查阅资料	不符合要求的,扣1分
每台冷却装置均应能正常运行,冷却系统风机和油泵投入时间均衡配置,避免某一单组损耗过度		不符合要求的,扣1分
十四、油泵		
高处安装潜油泵应加装防雨罩	现场检查	加装防雨罩,本体及二次电缆进线50 mm应被遮蔽,45°向下雨水不能直淋,不符合要求的,发现1次扣1.5分
不应存在异常振动和杂音		不符合要求的,发现1次扣1.5分
十五、油流继电器		
指示正确、无进水受潮、有必要的防雨措施、表盘内无渗漏、指针无抖动	现场检查	加装防雨罩,本体及二次电缆进线50 mm应被遮蔽,45°向下雨水不能直淋,不符合要求的,扣1.5分
冷却装置控制箱		
控制线电缆芯无外露,如采用多芯软铜线,应压接绝缘接线柱	现场检查	不符合要求的,扣1.5分

续表

评判小项	检查方式	评判原则
端子箱箱体接地、箱内二次接地、箱门与箱体的接地连接线连接良好	现场检查	未进行接地连接的，扣 1.5 分；采用接地线不规范或接地部位不可靠、锈蚀的，扣 0.5 分
驱潮装置和加热升温装置工作正常，温湿度整定值设定正确，按现场运行规程投退		不符合要求的，扣 1.5 分
箱门密封、内部封堵应良好，无进水、受潮、积灰、存在异物现象，如使用荧光灯管门灯，应加装防护罩		不符合要求的，扣 1.5 分
十六、气体继电器		
按规定周期进行气体继电器二次回路的绝缘电阻测量	查阅资料	未进行绝缘电阻测量的，扣 3 分；未按周期进行测量的，缺少一次扣 1 分，最高不得超过 3 分
应满足：二次回路绝缘电阻 ≥1 MΩ		试验数据不符合要求并未进行处理的，扣 3 分
防雨措施	现场检查	加装防雨罩，本体及二次电缆进线 50 mm 应被遮蔽，45°向下雨水不能直淋，不符合要求的，扣 1.5 分
观察窗应打开	现场检查	不符合要求的，扣 1.5 分
十七、压力释放阀		
按规定周期进行压力释放阀二次回路的绝缘电阻测量	查阅资料	未按周期进行绝缘电阻测量的，发现 1 次扣 3 分
应满足：二次回路绝缘电阻 ≥1 MΩ		试验数据不符合要求并未进行处理的，发现 1 次扣 3 分
本体压力释放阀导向管方向不应直喷巡视通道，威胁到运维人员的安全，并且不致喷入电缆沟、母线及其他设备上	现场检查	不符合要求的，发现 1 次扣 2 分
十八、油流速动继电器		
按规定周期进行油流速动继电器二次回路的绝缘电阻测量	查阅资料	未按周期进行绝缘电阻测量的，扣 3 分
应满足：二次回路绝缘电阻 ≥1 MΩ		试验数据不符合要求并未进行处理的，扣 3 分

续表

评判小项	检查方式	评判原则
防雨措施	现场检查	加装防雨罩,本体及二次电缆进线 50 mm 应被遮蔽,45°向下雨水不能直淋,不符合要求的,扣 1.5 分
十九、温度计		
按规定周期进行温度计二次回路的绝缘电阻测量	查阅资料	未按周期进行绝缘电阻测量的,发现 1 次扣 3 分
应满足:二次回路绝缘电阻≥1 MΩ		试验数据不符合要求并未进行处理的,发现 1 次扣 3 分
防雨措施	现场检查	温度计和温度探针加装防雨罩,本体及二次电缆进线 50 mm 应被遮蔽,45°向下雨水不能直淋,不符合要求的,发现 1 次扣 1 分
二十、突发压力继电器		
按规定周期进行突发压力继电器二次回路的绝缘电阻测量	查阅资料	未按周期进行绝缘电阻测量的,扣 1.5 分
应满足:二次回路绝缘电阻≥1 MΩ		试验数据不符合要求并未进行处理的,扣 1.5 分
防雨措施	现场检查	加装防雨罩,本体及二次电缆进线 50 mm 应被遮蔽,45°向下雨水不能直淋,不符合要求的,扣 1 分
二十一、二次回路		
二次电缆浪管不应有积水弯和高挂低用现象,否则应临时做好封堵并开排水孔	现场检查	不符合要求的,发现 1 次扣 1.5 分
二次元件标志应清晰、准确		不符合要求的,扣 1 分
二十二、变压器油中溶解气体在线监测装置		
应接入输变电状态监测系统	现场检查及查阅资料(无油中气体组分装置的不扣分)	不符合要求的,扣 1 分
应运行正常(无渗漏油、欠压、载气不足现象),数据上传准确。		不符合要求的,扣 1 分
取样周期应符合要求 一类变电站:4 小时; 二类变电站:12 小时; 三类变电站:24 小时		取样周期不符合要求的,扣 1.5 分
有运行维护记录		不符合要求的,扣 1 分
与变压器本体间的连接管路应做完善的保护措施		不符合要求的,扣 1 分

续表

评判小项	检查方式	评判原则
二十三、接地		
主要元器件应短路接地	现场检查	钟罩或桶体、储油柜、套管、升高座、有载开关、端子箱应短路接地，不符合要求的，发现1次扣1分
本体接地情况		应有二根在不同位置分别引向不同地点的水平接地体，不符合要求的，扣1分
中性点接地情况		应有两根与地网主网格的不同边连接的接地引下线，不符合要求的，扣1分，110~220kV不接地变压器的中性点过电压保护应采用棒间隙保护方式，间隙距离及避雷器参数配合应进行校核，不符合要求的，扣1分
接地装置导电截面无锈蚀现象；导电体接触良好		不符合要求的，发现1次扣0.5分
导轨式基础本体应固定良好		不符合要求的，扣1分

工作中的注意点：

（1）注意区分当前分接开关类型和厂家，不同厂家的机构，内部元件布局和电气回路会有所区别，制定的作业方案和消缺思路应根据实际情况变化（见图2-59至2-61）。

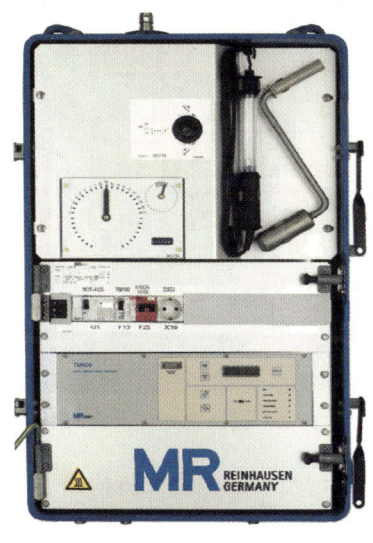

图2-59 MR公司MA7型机构（带面板）

 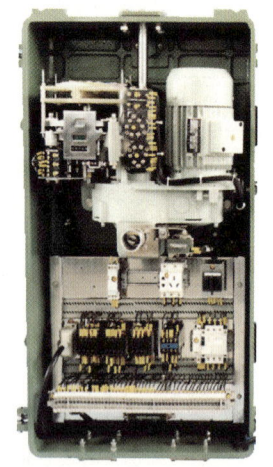

图 2-60 华明公司 MA7 型机构（已取面板）　　图 2-61 长征公司 DCJ10 型机构（已取面板）

（2）注意机构内部元件分布，绝大多数的机构组成模块是相同的，从图 2-62 可以看出，虽机构内元件排布位置有所区别，但元件类型几乎相同。一般来说，对于分接开关机构（无论 M 型或 V 型）主体功能应包括三类：动力部分、过程控制部分、远控接线部分。

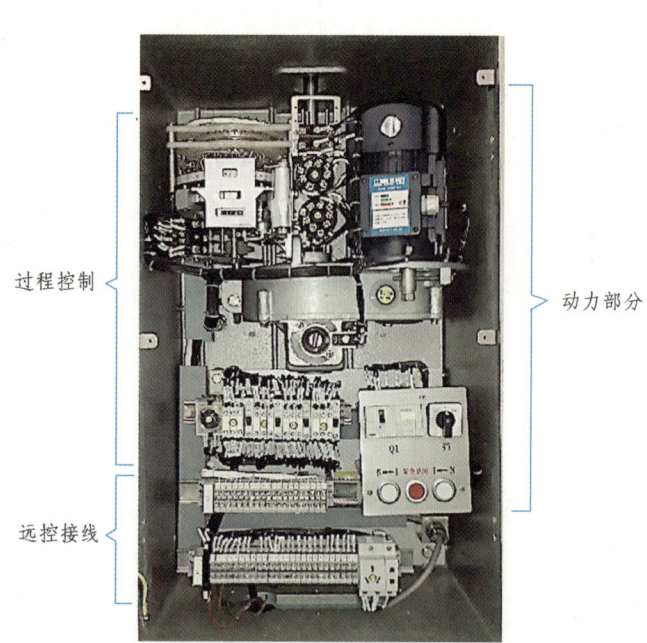

图 2-62 机构内功能分布模块

（3）完整的挡位确认应该是远控显示挡位、机构内窗口显示挡位、开关顶部观察窗口显示挡位三者一致，严格来说"远控"操作和显示端的挡位仅仅是"就地"操作时各种接触器、行程开关反馈的"信号""表现"，在安全条件允许情况下，只要机构显示挡位和开关顶部挡位两者显示一致即能认为就地挡位配合正确,如图 2-63,远控显示不正确多数时候是电气信号或接线问题。

（a）顶部显示挡位　　　　　　　　　　（b）机构显示挡位

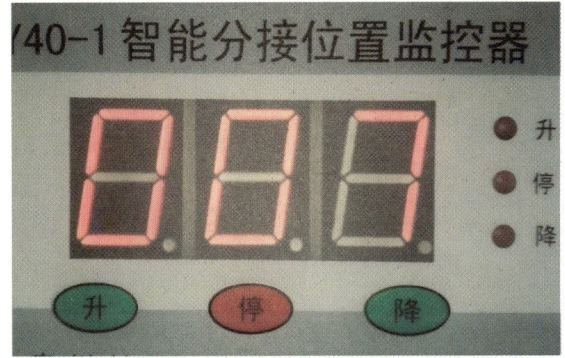

（c）远控显示挡位

图 2-63　挡位显示（此时都是 7 挡）

（4）加热电阻位置一般在机构内侧,有的机构在机构内侧壁上,如图 2-64。

（5）齿轮涂抹润滑脂位置如图 2-65。

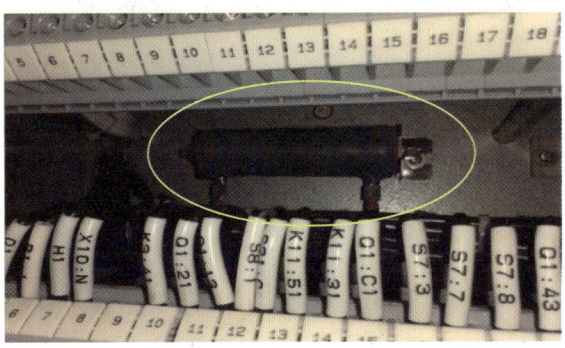

图 2-64　加热电阻位置

项目二　变压器检修　141

（a）涂抹位置　　　　　　　　　　　　（b）涂抹状态

图 2-65　齿轮涂抹润滑脂位置

（6）进行电气连接检查时，不仅仅要查看接线的牢固度，还应注意其配套垫片或弹垫是否完整，如图 2-66、图 2-67。

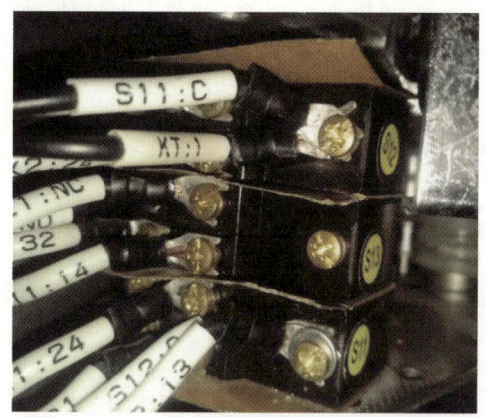

图 2-66　行程开关所有接线端垫片和弹垫均丢失

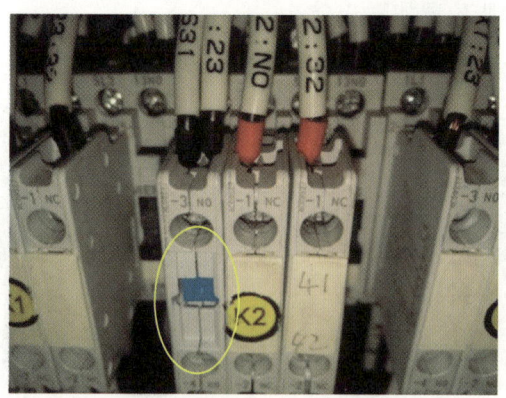

图 2-67　接触器手动弹跳触头盖板丢失

（7）进行连接校验（挡位调整）的方法不仅用于正常挡位功能的验证，同样适用于滑档（挡位有偏差）的调整。具体方法如下：

① 在机械配合良好的情况下：在任意位置（V型开关建议在整定挡）的正常机构显示挡位处（即红色标记线在窗口中间），手动升挡一次，直至下一个正常显示挡位。期间摇柄转动圈数记为 X，再次降挡回开始的挡位，期间转动圈数记为 Y，完好的机械配合体现为$|X-Y|\approx 0$。但实际上由于机构传动带动的齿轮、联杆、传动盒、传动轴各种力传递部件较多，很难实现$|X-Y|\approx 0$的情况，所以一般工作中$|X-Y|\leq 1$，即升降挡各一次，圈数之差不大于1圈，即认为配合良好。

② 在机械配合有偏差的情况下：按照"①"同样计数 X 和 Y，取下传动联杆，用操作手柄往 X 和 Y 代数值大的方向空转，转动圈数为$|X-Y|\div 2=N$。再连上传动联杆，按照"①"的步骤验证，并重复"②"的步骤，直至$|X-Y|\leq 1$为止。

③ 若对分接开关切换状态比较了解（见"相关理论知识"），则有更简易的方法进行。由于分接开关进行切换一定要待开关触头完成切换后才能对机构电机断电，否则触头会悬浮放电，直至烧损，而分接开关每个挡位的触头切换过程是一样的，因此每一档的变化在行程上的差异主要是"触头完成转换"至"机构电机断电"这段时间，这段时间我们成为"惯性行程"（空行程），即 X 或 Y 的圈数各自组成都是"开关切换过程+惯性行程"，$|X-Y|=|$"开关切换过程（升）+惯性行程（升）"－"开关切换过程（降）+惯性行程（降）"，根据上面说明的每个挡位触头切换过程一致，因此"开关切换过程（升）"="开关切换过程（降）"，则$|X-Y|=$升降挡惯性行程之差。

惯性行程从触头切换完成的时候开始计数，即听到切换完成的声音，或从手柄上感受到轻微震动（该震动是快速机构弹簧释能）后开始计数。则按照此方法，升降挡惯性行程计数就大大减少。

Task 2 Maintenance of Mechanism Box of M-Type On-Load Tap Changer

2.1 Work Tasks

(1) Completion of the items and contents of the maintenance inspection of the M-type on-load tap changer mechanism.

(2) Completion of the commissioning and function verification of the M-type on-load tap changer mechanism gearing.

(3) Being familiar with the control function reading of the secondary circuit diagram of the on-load tap changer.

2.2 References

(1) *Electric Power Safety Working Regulations (Power Transformation) of State Grid Corporation of China* (Q/GDW1799.1—2013).

(2) Specialized teaching materials for the training of vocational competence of production skill personnel of the state grid of China: *Transformer Maintenance*.

(3) *Substation Maintenance Management Regulations of State Grid Corporation of China (Trial)—Volume 1: Detailed Rules for Maintenance of Oil Immersed Transformers (Reactors)*.

(4) *Regulations of State Grid Corporation of China on Management of Substation Operation and Maintenance (Trial)—Volume 1: Detailed Rules for Operation and Maintenance of Oil Immersed Transformers (Reactors)*.

(5) *Maintenance Guide for Power Transformer* (DL/T 573—2010).

(6) *Guide for the Operation and Maintenance of Tap Changers in the Power Transformer* (DL/T 574—2010).

(7) *Tap Changers—Part 1: Performance Requirements and Test Methods* (GB/T 10230.1—2007).

(8) *Tap Changers—Part 2: Application Guide* (GB/T 10230.2—2007).

2.3 Work Requirements

(1) In accordance with the requirements of the *Guide for the Operation and Maintenance of On-Load Tap Changers*, the mechanism box shall be subject to maintenance and inspection once a year.

(2) The general use of the mechanism is at an altitude of up to 2,000 m above sea level. If this is exceeded or a special type is customized, special instructions are required.

(3) The ambient temperature is not higher than +55°C and not lower than −25 °C.

(4) The deviation of perpendicularity of the mechanism from the ground does not exceed 5%.

(5) The operation site of the mechanism is free from serious dust and other explosive and corrosive gases.

2.4 Preparation for Work

1. Investigation of the basic condition of the equipment

(1) The transformer body and the vicinity of the on-load tap changer should be free from oil leakage defects. If any, the oil leakage should be dealt with first.

(2) The displayed gear of the on-load tap changer mechanism before the work, the displayed gear of the remote control, and the actual gear displayed in the window of the on-load tap changer core should be recorded, and all three above-mentioned should be

consistent. If not coincide, it is necessary to determine where the defects appear and how they are handled during the work.

(3) The counter value in the current gear position of the on-load tap changer mechanism shall be recorded.

(4) The mechanism drive links should be visually inspected for misalignment and no parts detachment.

(5) The mechanism drive links shall be visually inspected for obvious deformation, without extensive corrosion. If any, they need to be replaced.

(6) The mechanism nameplate should be checked, paying special attention to the mechanism model and manufacturer.

2. Hazards and preventive and control measures

(1) Hazard: electric shock injury. Preventive and control measures are as follows:

① When the main transformer is not de-energized, it is not possible to climb up to the main transformer to observe the gears for mechanism maintenance and inspection, but only to observe the gears from the remote control end.

② The location of the temporary power supply should be determined, and the cable setting-out should be bypassed as far as possible from outside the working area.

③ Use a multimeter to verify that the power supply side of the mechanism is disconnected before operation.

④ If the team has its own switch box, attention should be paid to checking the wiring and grounding, and confirming that the air switch is in the disconnected position. When energizing, first close the temporary power supply air switch, then close the switch box air switch, and finally close the mechanism box air switch.

⑤ All working cable wiring should be checked for insulation integrity. If there is any rupture, use yellow insulating and waterproof tape to wrap it in several layers, and pad dry wooden boards underneath.

(2) Hazard: slipping and falling. Preventive and control measures are as follows:

① Staff should properly wear non-slip shoes.

② If there is oil dirt on the floor, the soles of shoes need to be cleaned before entering or leaving the site and going up or down the ladder.

③ In order to prevent the parts of the mechanism from falling into the pool and not being able to find them, the floor of the fireproof pool under the mechanism box should be covered with a white cloth or serpentine cloth in an appropriate amount according to the scope of operation. Try not to use tarpaulin. If a part falls on a tarpaulin, it may bounce off and fall into the fireproof pool. It is easier to slip and fall when walking on the oil-soaked tarpaulin.

④ When setting up a ladder or insulated high stool in the fireproof pool, it must be supported by a specially-assigned person.

3. Work tools and material selection

Work tools and consumables for maintenance work of M-type on-load tap changer mechanismbox are shown in Table 2-6.

Table 2-6　Work Tools and Consumables for Maintenance Work of M-Type On-Load Tap Changer Mechanism Box

Category	Name	Specification and model	Quantity	Remarks
Specialized tools	Safety helmet		3	
	Insulating gloves	10 kV	1 pair	
	Megohmmeter (or electronic)	500 V or 1,000 V	1 set	
	Multimeter		1 set	
	Distribution box	Portable	1 set	
Personal tools	Slotted screwdriver		1	Circuit terminal
	Cross-head screwdriver		1	For panel removal
	Open spanner	8-10,10-12,17-19	2 each	For removing connections
	Adjustable spanner	300 mm	1	For auxiliary use with an adjustable spanner
	Tool shelf		1	
Consumables	Abrasive paper	#0 or #1	3 each	Corroded portion
	Grease	Lithium-based grease	1 package	Calcium-based grease should not be used for exposed gears in the mechanism
	Mane brush		1	
	White silk cloth		A number of square meters	
	White cloth		A number of square meters	

4. Division of labor among operators

See Table 2-7 for the division of labor among maintainers for M-type on-load tap changer mechanism box.

Table 2-7 Division of Labor among Maintainers for M-Type On-Load Tap Changer Mechanism Box

S/N	Job	Quantity	Job responsibilities
1	Person in charge of the work (guardian)	1	He/she is responsible for the division of labor for the work, site inspection before the work, the development of the work plan, the preparation of work tickets, the handling of work permits, the convening of pre- and post-shift meetings, the safety supervision during the work, the handling of emergencies during the work, the supervision of the quality of the work, and the summarization of the work
2	Operator	1	He/she is responsible for maintenance work, result verification, report, and record filling
3	Auxiliary operator	1	He/she assists with work requiring cooperation, and written documentation of operations

2.5 Working Procedures

Mechanism box maintenance procedures of M-type on-load tap changer see Table 2-8.

Table. 2-8 Mechanism Box Maintenance Procedures of M-Type On-Load Tap Changer

S/N	Scope of work	Operation procedures and standards	Technical and safety quality requirements
1	Preliminary work	(1) Work tickets should be filled out and issued in a standardized manner, and work ticket procedures should be properly carried out. (2) Site investigation must be carried out by two persons. The name of the transformer corresponding to the tap changer, the type of tap changer, and the current operating gear shall be checked on the site. (3) Safety fences shall be properly installed with a sign board hung	(1) Safety helmets, work clothes, and labor protection gloves should be worn properly. (2) Do not change gear without disconnecting the tap changer links or without de-energizing the main transformer intervals. (3) Before operating the tap changer, the current operating gears of the mechanism and the remote control monitor should be recorded. (4) It is strictly prohibited for unrelated personnel and vehicles to enter the site

Continued

S/N	Scope of work	Operation procedures and standards	Technical and safety quality requirements
2	Inspection of work tools	(1) Prepare and neatly arrange the work tools according to the requirements, and check the appearance, performance and test certificate of the work tools, without omission. (2) An inflation test should be done on insulating gloves, using the correct method. (3) The megohmmeter should be tested for short circuits and open circuits and zeroed. (4) The multimeter should be checked for an on-off state, and the display and beeping alarm should be intact	(1) Appearance checking of work tools should be qualified, with no damage, deformation, or malfunction, and the certificate of conformity is within the validity period. (2) Test for insulating gloves, megohmmeter, and multimeter should be qualified
3	Data recording	(1) The nameplate parameters of the mechanism shall be recorded. (2) The current gear A displayed by the mechanism shall be recorded. (3) The remote control (LCD or electronic display) gear B shall be recorded. (4) The current counter number N of the mechanism should be recorded. (5) A and B should be consistent	(1) In principle, it is not permitted to open the mechanism box during the recording process. (2) If the sight glass of the mechanism is badly deteriorated and the observation is not clear, the box may be opened for recording. At this time, it is prohibited to touch any internal components
4	Power outage (case 1)	(1) It should be confirmed that the main transformer interval subject to maintenance has been de-energized, and safety measures have been completed. Permission is granted to begin work. (2) If only the main transformer is de-energized, it should be confirmed that the HV and LV side leads have been disconnected, and the grounding wire has been set. It should be confirmed that the main transformer housing, neutral point, and core are well grounded, and electricity testing is carried out for confirmation	(1) No outages shall be scheduled. (2) The names and numbers of all equipment subject to the outage interval shall be reconfirmed as accurate

Continued

S/N	Scope of work	Operation procedures and standards	Technical and safety quality requirements
5	No power outage (case 2)	(1) Electricity testing should be done on the tap changer links, which should not be energized. (2) The drive link connected to the mechanism should be disconnected and completely removed wearing insulating gloves	(1) It is strictly prohibited to climb up the transformer to work. (2) The "No Climbing" sign board on the transformer ladder should be obvious. (3) The transformer ladder lock shall be locked
6	Mechanism power supply check	(1) It should be confirmed that the air switch in the centralized control box of the secondary power supply of the mechanism has been disconnected, and electricity testing is carried out for confirmation. (2) If the secondary power supply cable has been removed, it should be confirmed that the wiring of the self-contained distribution box or the temporary power supply at the site is intact, and the protection function of the air switch is intact	(1) A small sign board should be hung on the corresponding air switch in the centralized control box to indicate that "No Closing, Work in Progress". (2) The insulating coat of the temporary power supply line should be free from rupture. If any, use yellow insulating tape to wrap at least two layers back and forth. It should be located in an area where there is little movement of people to avoid stepping on it. (3) Call-out instructions must be obtained before the mechanism can perform a power feed operation
7	Disconnection confirmation of the power supply of the secondary circuit	(1) Make sure that the secondary circuit air switch in the box is disconnected. (2) The control knob should be adjusted from "Remote" to "Stop". (3) Disconnect the fuses from the terminal strip and remove them to check for fuse on-off state	(1) It is forbidden to set the control knob to the "Local" position before electric commissioning. (2) If the fuse is hit, or if the fuse may break after being dropped, remove the fuse and check it before putting it back in place. If not, wrap it with paper or cloth and place it in a place where it will not collide directly with other hard objects

Continued

S/N	Scope of work	Operation procedures and standards	Technical and safety quality requirements
8	Removing the masking panel	(1) The masking panel should be removed after opening the box. (2) The operating handle should be checked for proper suspension on the inside of the panel or mechanism door	Panels should have various markings on them. After the panel has been removed, it should be placed in a place where it will not collide with rigid objects to avoid paint shedding and scratches to the markings
9	Inspection of mechanism seals	(1) It should be checked that the sealing rubber strips of the door of the mechanism are in the slot, without falling off, and should have good elasticity when pressed. (2) It should be checked that the blocking sealing of the power and control cable access holes at the bottom of the mechanism is intact and has not fully hardened and become brittle	It is prohibited to check the access hole blocking sealing by hand without confirming that the power supply to the mechanism is disconnected
10	Cleaning	(1) Dust inside of the mechanism and the components should be cleaned with a brush and a white silk cloth. (2) After cleaning, make sure that no bristles get stuck in the gaps of the components	(1) The brushes should be moderately soft and hard, to avoid too hard bristles stuck in the screw crimping surface of the components or other gaps. (2) Brushes should not be oiled or greased
11	Inspection of gearbox	(1) There should be no uneven jamming and no continuous or intermittent crashing sounds when raising or lowering each gear with the operating handle. (2) Wipe the sealing area at the bottom of the gearbox with white silk cloth. There shall be no water or oil seepage. (3) The gearbox shall have a factory marked "No Oil Injection" sign. (4) Two persons cooperate. One person turns the handle at a uniform speed, and the other uses a brush to smear the grease and evenly applies it to the exposed gears above the gearbox	(1) The gearbox is welded and sealed along the bottom, and it is strictly prohibited for staff to disassemble and inspect on their own. (2) It is forbidden to inject lubricating oil into the gearbox

Continued

S/N	Scope of work	Operation procedures and standards	Technical and safety quality requirements
12	Check electrical connection	(1) Visually inspect and manually cooperate to inspect all electrical components within the mechanism for good wiring. There shall be no looseness, and no obvious discharge marks on the wiring terminals. (2) Check whether all terminals are numbered well and clearly, and take photos of defects and unclear areas and record them. (2) Check the operation of all travel switches and the sound of contact connections. (3) Check that all contactor contacts bounce normally	It is forbidden to power on the mechanism during inspection
13	Measure the heating resistor	Use a multimeter to measure the heating resistor resistance, which shall be consistent with the value marked on the resistor. For example, $1,000 \times (1 \pm 2\%)\ \Omega$	Disconnect the external wiring of the heating resistor during measurement and immediately restore it after measurement
14	Measure the circuit insulation resistance	(1) Disconnect the external power cable of the mechanism. (2) Close the terminal block fuse. (3) Verify that the grounding on the inside of the mechanism door is intact. (4) Use a 500 V or 1,000 V megohmmeter to measure the circuit insulation resistance, which shall not be less than $1\ M\Omega$. (5) Disconnect the fuse as needed, and measure the insulation resistance of the main circuit with the motor and the insulation resistance of the control circuit separately	(1) It is prohibited to power on the mechanism during the measurement. (2) The person who cooperates with the measurement shall wear insulating gloves and can not take them off until the discharge is completed. (3) When there is no defect in the motor, it is not recommended to open the cover for the motor insulation test

项目二　变压器检修

Continued

S/N	Scope of work	Operation procedures and standards	Technical and safety quality requirements
15	Inspection of gear rotation process	In a cycle of manually raising and lowering gears at any position, if the four processes of "selector tripping—selector closing—the changeover switch completing the changeover—motor travel switch bouncing and powering off" can be experienced from the degree of jamming and sound of the handle, it means that the mechanical action during the rotation process is normal. (For specific methods, see the follow-up related theoretical knowledge)	It is forbidden to power on the mechanism during inspection
16	Gear zeroing	Manually operate the gear to the set gear and check that the difference in the number of turns between two gears is zero. Otherwise, coupling verification shall be performed. (For specific methods, see the follow-up related theoretical knowledge)	It is forbidden to power on the mechanism during inspection
17	Check the mechanical limit	(1) Manually operate the gears to the highest gear and gear 1, and continue shaking until the mechanical limit position is reached, until it cannot be turned. (2) Check that the locking connection of the limit screw is intact	It is forbidden to power on the mechanism during inspection
18	Mechanism power-on	(1) Confirm that the current gear is in the completed switching state, that is, the red line indicates in the middle of the window. (2) Confirm that the travel switch cam of the circuit power outage has bounced open. (3) Confirm that the air switch is in a disconnected state. (4) Issue a power-on call command and receive a power-on operation response. (5) Confirm that the air switch indicator light is on. (6) Close the air switch and turn the control knob from "Stop" to "Local"	(1) Do not power on until the work is confirmed to be completed. (2) Power-on requires calling and receiving a response. (3) If the air switch disconnected state indicator light is not on, verify it with a multimeter and confirm the power loss before checking

Continued

S/N	Scope of work	Operation procedures and standards	Technical and safety quality requirements
19	Check the electrical limit	(1) Turn to the highest gear and the lowest gear via electric operation. After the automatic stop, continue to press the Up or Down button, the highest gear will not continue to rise, and the lowest gear will not continue to fall. (2) During the lifting process, if the stop button is not pressed, there will be no power outage due to air switch tripping, otherwise the fault shall be eliminated	Before passing the mechanical limit inspection, it is prohibited to directly power on for electrical limit inspection. Otherwise, if there is a deviation in the gear position, it may cause the connecting rod joint connected to the core to be broken
20	Check lifting logic and process	(1) With the limit intact, change the electric tap by a cycle. (2) During the cycle, check that the start button is accurate and reliable, and that the gear up and down logic is correct. (3) During the cycle, check the emergency stop button at any position, which shall operate reliably. (4) For the mechanism that has the function of override node, check whether the override node works reliably, that is, the switch with 9a, 9b, and 9c gears will not stop at 9a and 9c during the electric cycle, but will stop normally at other gears. (5) Check that when each gear automatically stops, the red line in the window is near the middle position	(1) The interval of each electric operation is 5 seconds, and then continue to rise and fall after the contactor coil is discharged. (2) If the operating handle is used during electric operation, confirm that the handle has been removed during the next electric operation to avoid the handle being thrown out and hurting people in the event of a short circuit
21	Check the counter count	According to the data record, at this time, the increased count of the counter shall be consistent with the number of gear switching in this operation	The counter counting error or failure will affect the cycle of mechanism minor repair and switch major repair determined according to the switching times

Continued

S/N	Scope of work	Operation procedures and standards	Technical and safety quality requirements
22	Check the function of time relay	(1) On the premise that the electric cycle can stop automatically in each gear correctly, adjust the time relay time to less than 5 seconds, and then conduct electric operation again, the mechanism can stop due to air switch tripping before completing one gear rotation. (2) Time required for recovering time relay: generally set as 12.5 seconds for mechanisms with override node function and 7.5 seconds for mechanisms without override node function	(1) For mechanisms without override node function, the time set for time relay shall be greater than the electric switching time of one gear and less than the electric switching time of two gears. (2) For mechanisms with override node function, the time set for time relay shall be greater than the electric switching time of at least two gears and less than the electric switching time of three gears
23	Check manual/electric interlock	When the power is on and the air switch is closed, insert the operating handle, the air switch will trip immediately, and during the insertion of the handle, the air switch cannot be closed	Once this function is invalid, the electric gear may be touched by mistake during operation, causing the operating handle to be thrown out directly, resulting in personal injury or damage to object
24	Remote control function check	(1) Turn the control knob from "Local" to "Remote". (2) Sing and operate an electric cycle from the remote control end. (3) Check that when each gear automatically stops, the red line in the window is near the middle position. (4) When each gear is automatically stopped, the remote control display gear is consistent with the window display gear in the mechanism	If the remote control function fails or the display is incomplete, the operation and maintenance and background monitoring will be affected. If the mechanism and switch body functions are intact, the local operation of the mechanism and switch switching process will not be affected

Continued

S/N	Scope of work	Operation procedures and standards	Technical and safety quality requirements
25	Restore the gears	(1) According to data records, either electrically or manually restore to the initial gear. (2) Confirm that the remote control display gear is consistent with the window display gear in the mechanism	The complete determination shall be that the remote control display gear, the window display gear in the mechanism and the observation window display gear on the top of the switch are the same, but the top gear of the switch cannot be observed without power failure. Therefore, it is required to carry out operations of a cycle during the "check lifting logic and process", so it is possible to find out whether there is any inconsistency between the top gear of the switch and the gear of the mechanism
26	Disconnect the power supply	Disconnect the maintenance power or secondary power in the central control box	
27	Restore link connection	(1) Restore the drive link connection that was removed. (2) Confirm that the link bolts are complete, and the tightening pieces of the connecting bolts of the upper and lower rod parts have been pressed tightly	In case of using insulated high stool or herringbone ladder when restoring the connection of link, pay attention to monitoring and forbid the body to probe into the area without power failure on the upper part of transformer
28	Recover the shadow panel	(1) Confirm once again that there are no missing parts and tools in the mechanism and on the top of the mechanism. (2) Recover the shadow panel. (3) Restore the suspension position of the operating handle	(1) Pay attention to the adhesion and missing parts between the sealing adhesive and the inner side of the box bottom. (2) The operating handle shall not be placed directly across the box bottom space

The evaluation of appearance status of transformer proper and accessories see Table 2-9.

Table 2-9 Evaluation of Appearance Status of Transformer Proper and Accessories

Evaluation Item	Means of Inspection	Evaluation Principle
Ⅰ. Oil tank appearance		
The nameplate of the equipment is complete, clear, and identifiable	On-site inspection and data consulting (including the "oil temperature - oil level" curve of 110 kV and above proper)	Deduct 0.5 point if there is no equipment nameplate; Deduct 0.2 point if the nameplate is more than 1/3 rusted or unclear; Deduct 0.2 point if the installation position of the nameplate is unreasonable and cannot be accurately viewed on site. No points will be deducted if the nameplate is located high and cannot be accurately viewed, but photos can be provided
The operation number sign is clear, correct, and identifiable.	On-site inspection (including radiator number)	Deduct 0.5 point if there is no operation number and double numbers of equipment are incomplete and cannot be correctly identified; Deduct 0.2 point if the operation number sign is more than 1/3 unclear
The phase sequence sign is clear, correct, and identifiable.	On-site inspection	Deduct 0.5 point if the phase sequence sign cannot be correctly identified; Deduct 0.2 point if the phase sequence sign is more than 1/3 unclear
Ⅱ. Thermometer		
Thermometer indication results	On-site inspection and data consulting (including winding and oil level thermometer)	The temperature indicated by the on-site thermometer shall be roughly consistent with that displayed by the temperature display device in the control room and displayed by the monitoring system, with the maximum error not exceeding 5K, and the winding temperature shall not be less than the oil temperature. Deduct 1.5 points for any item found to be non-conforming
Interphase temperature difference		For a single-phase transformer, the temperature difference between different phases of the same group of equipment shall be less than 10K. Deduct 1.5 points for any item found to be non-conforming

Continued

Evaluation Item	Means of Inspection	Evaluation Principle
Fixation of thermometer lead wire	On-site inspection and data consulting (including winding and oil level thermometer)	The thermometer lead wire shall be well fixed, and measures such as backflow bends shall be taken to prevent rainwater from flowing back. The radius of the reel shall not be less than 50 mm. Deduct 1.5 points for any item found to be non-conforming
Rainproof measures (excluding boxin indoors)		Rain cover shall be installed. The proper and secondary incoming cable shall be covered by 50 mm, 45°downward, and can not be directly drenched by rainwater. Deduct 1.5 points for any item found to be non-conforming
Thermometer dial indication		The thermometer shall indicate clearly without any water ingress, and the historical highest temperature indication shall be correct. The setting position of the dial shall be basically consistent with the setting on the setting sheet. Deduct 1.5 points for any item found to be non-conforming
Temperature record		Record the temperature, corresponding load, and ambient temperature according to the comprehensive routine inspection cycle of the substation. Deduct 1.5 points for any item found to be non-conforming
Ⅲ. Oil conservator		
The oil level indication should comply with the "oil temperature—oil level curve"	On-site inspection	When the supplier has stipulated standards, follow such standards. If there are no such standards, the oil level shall be within the range of -10% to 10% of the standard curve. Deduct 1.5 points for any item found to be non-conforming
The oil level meter shall be equipped with a rain cover (outdoor transformer), there shall be no oil dirt inside, the oil level shall be clearly visible, and can be read locally during operation		The oil level meter sight glass shall be clearly identified. The oil level meter shall be equipped with a rain cover. The proper and secondary incoming cable shall be covered by 50 mm, 45° downward, and can not be directly drenched by rainwater. Deduct 1.5 points for any item found to be non-conforming

Continued

Evaluation Item	Means of Inspection	Evaluation Principle
Ⅳ. Moisture absorber		
The oil seal of the glass cup shall be intact and can provide long-term breathing function	On-site inspection	The glass cup shall be undamaged and the seal shall be intact without water intake. In the breathing out or breathing in state, the inner or outer oil level shall be higher than the breathing tube mouth, and the oil level in the oil cup shall not be too high. Deduct 1.5 points for any item found to be non-conforming
Use color-changing moisture absorbent, which shall be canned to 1/6—1/5 of the top, with no more than 2/3 of the moisture absorbent affected by moisture		Deduct 1.5 points if no color-changing moisture absorbent is used; Deduct 1 point if the dose of moisture absorbent does not meet the requirements
The upper part of the moisture absorbent shall not be moistened with oil or be crumbled or powdered.		Deduct 1.5 points for any item found to be non-conforming
The connecting pipe shall be clean as a whole, without blockage or rust, and the position of the bypass valve with the oil conservator shall be correct		Deduct 1.5 points for any item found to be non-conforming
Power supply of the maintenance-free moisture absorber shall be in good condition. The normal starting value of the heater is less than RH 60% or according to the manufacturer's regulations		Deduct 1.5 points for any item found to be non-conforming
Ⅴ. Valves		
Valves must be in the closed and open positions according to actual needs	On-site inspection	Deduct 1.5 points for any item found to be non-conforming
The signs indicating the opening and closing positions shall be clear and correct		Deduct 1.5 points for any item found to be non-conforming

Continued

Evaluation Item	Means of Inspection	Evaluation Principle
Ⅵ. Noise		
There shall be no abnormal noise during operation	On-site inspection	Deduct 1.5 points for any item found to be non-conforming
Ⅶ. Rusting		
There is no obvious rust on the metal parts of the proper and assemblies	On-site inspection	The rusting area shall not be larger than 400 mm^2. Deduct 0.1 point for any item found to be non-conforming. About bolt rusting: deduct 0.1 point for conductive part bolt rusting and deduct 0.02 point for non-conductive part bolt rusting. Deduct 0.5 point at most for rust on all parts
Ⅷ. Leakage		
There is no oil leakage in the negative pressure area of the proper and assemblies	On-site inspection	The negative pressure area includes the oil inlet of the oil-submerged pump, the oil cylinder to the oil pump inlet of the on-load switch on-line oil filter, the top of the oil storage tank, the bushing oil storage tank, and other areas close to or higher than the oil level. Deduct 2 points for any item found to be non-conforming
There is no oil leakage in the positive pressure area of the proper and assemblies		Deduct 1 point for any item found to be non-conforming and of serious and above defect. Deduct 0.5 point for any item found to be non-conforming and of common defect
Ⅸ. Proper terminal box		
The core of the control line cable is not exposed. If multi-core soft copper wire is used, insulating terminals shall be crimped	On-site inspection	Deduct 1.5 points for any item found to be non-conforming
The grounding of the terminal box body, secondary grounding inside the box, and the grounding connection wire between the box door and the body are well connected		Deduct 1.5 points if the grounding connection is not made; Deduct 0.5 point if the use of grounding wires is not standardized or the grounding parts are unreliable or rusted

Continued

Evaluation Item	Means of Inspection	Evaluation Principle
The moisture-proof device and heating device work normally, the temperature and humidity set value setting is mobilized/demobilized according to the on-site operation regulations. If the heater is used, the distance between its position and the elements, cables, and wires shall be greater than 50 mm	On-site inspection	Deduct 1.5 points for any item found to be non-conforming
The moisture-proof device and heating device work normally, the temperature and humidity set value setting and the moisture-proof device and heating device are all mobilized/demobilized according to the on-site operation regulations		Deduct 1.5 points for any item found to be non-conforming
The door sealing and internal plugging shall be good, without water intake, moisture, dust accumulation and foreign matters. If the door lamp used is a fluorescent lamp, a protective shield shall be installed		Deduct 1.5 points for any item found to be non-conforming
X. Bushing		
Oil level indication	On-site inspection	Oil level or gas pressure is normal, oil level meter or pressure gauge on-site indication shall be clear, easy to observe (with the help of long-focus camera or telescope). For the oil bushing installed vertically, the oil level shall be above 1/2 (non-full oil level); for the oil bushing installed with an inclination of 15°, the oil level shall be higher than 2/3 and up to full oil level. Deduct 1.5 points for any item found to be non-conforming

Continued

Evaluation Item	Means of Inspection	Evaluation Principle
Insulators shall be free from damage or cracking, flanges shall be free from cracking, and single enamel lack shall not exceed 25 mm^2. The total area of impurities on the enamel shall not exceed 100 mm^2		Deduct 1.5 points for any item found to be non-conforming
No discharge, severe corona, and electrical erosion		Deduct 1.5 points for any item found to be non-conforming
Hoops and wire clamps shall be crack-free	On-site inspection	Deduct 1.5 points for any item found to be non-conforming
For outdoor crimping equipment wire clamps, when installed upward 30°—90° (ϕ400 and above), a drainage hole with a diameter of 6—8 mm shall be drilled. The wire clamp shall not be copper-aluminum butt joint transition wire clamp, and a replacement plan shall be developed for copper-aluminum butt joint transition wire clamp. The bushing connecting terminals (hoop clamps) of transformers with voltage classes of 110 (66) kV and above shall be hot extruded from T2 pure copper material. It is forbidden to use brass or cast hoop clamps	On-site inspection and data consulting	Deduct 1.5 points for any item found to be non-conforming
The lead shall not be loose, twisted, or broken	On-site inspection	Deduct 1.5 points for any item found to be non-conforming
The bushing proper shall not be abnormal in terms of temperature	On-site inspection	Deduct 1.5 points for any item found to be non-conforming
There shall be no joint heating		Deduct 1.5 points for any item found to be non-conforming

Continued

Evaluation Item	Means of Inspection	Evaluation Principle
The actual specific creepage distance matches the grade of the polluted area	Data consulting	Deduct 1.5 points for any item found to be non-conforming
The adhesion between the metal flange and the pouring part of the porcelain insulator shall be firm, the waterproof glue shall be intact, the sandblasting shall be uniform, and there shall be no obvious electrical erosion	On-site inspection	Deduct 1.5 points for any item found to be non-conforming
XI. On load tap changer		
The indication of oil level shall be clear, accurate, and easy to observe	On-site inspection	Deduct 1.5 points for any item found to be non-conforming
The oil level shall be normal and shall not be too high or too low		Deduct 1.5 points for any item found to be non-conforming
The oil seal of the glass cup shall be intact and can provide long-term breathing function		The glass cup shall be undamaged and the seal shall be intact without water intake. In the breathing out or breathing in state, the inner or outer oil level shall be higher than the breathing tube mouth. Deduct 1.5 points for any item found to be non-conforming
Use color-changing moisture absorbent, which shall be canned to 1/6—1/5 of the top, with no more than 2/3 of the moisture absorbent affected by moisture	On-site inspection	Deduct 1.5 points if no color-changing moisture absorbent is used; Deduct 1 point if the dose of moisture absorbent does not meet the requirements
The hygroscopic agent should not be discolored from top to bottom and the upper part should not be moistened with oil or be crumbled or powdered		Deduct 1.5 points for any item found to be non-conforming
Power supply of the maintenance-free moisture absorber should be in good condition. The normal starting value of the heater is less than RH 60% or according to the manufacturer's regulations		Deduct 1.5 points for any item found to be non-conforming

Continued

Evaluation Item	Means of Inspection	Evaluation Principle
The switching characteristic test should be carried out after handover and hoisting inspection	Data consulting	Deduct 3 points for each failure to carry out the switching characteristic test
Tap changer proper should be equipped with mechanical limit devices. Mechanical limits and electrical locks should be provided in the mechanism box, and measures should be taken to deal with the defects of sliding gears and galloping	On-site inspection and data consulting	Deduct 2 points if there is no limit device and electrical locks; Deduct 1 point if no measures are taken to deal with the defects of sliding gears and galloping
The gear tap instruction should be consistent with the control room, proper and operating mechanism	On-site inspection	Deduct 1.5 points if the actual gear indication on site is inconsistent with other indications
The filter element of the online oil filter should be replaced according to the provisions of oil pressure and operating time	On-site inspection and data consulting	Deduct 1.5 points for failure to replace the filer element according to the period; Deduct 1 point if there is no record of replacement and overhaul
The pressure of the online oil filter should be normal and free of oil leakage		Deduct 1.5 points if the pressure is abnormal; Deduct 1 point for oil leakage from meter joints
The online oil filter should have antifreeze measures under extremely cold conditions, and there should be a self-starting heating and insulation device		"Extremely cold conditions" refer to the situation that the oil temperature in the tap changer is lower than 0°C and below, and deduct 1.5 points if it does not meet the requirements
Tests should be carried out on the basis of evaluation results	Data consulting	Deduct 3 points if the test is not carried out on time in accordance with the *Regulations of Condition-Based Maintenance & Test for Electric Equipment*
Oil pressure resistance should be qualified	Access to information (last test report)	Deduct 3 points if the test data does not meet the requirements

Continued

Evaluation Item	Means of Inspection	Evaluation Principle
XII. No load tap changer		
The tap gear indication of the non-excited tap changer with gear output should be consistent with the control room	On-site inspection	Deduct 1.5 points if the actual gear indication on site is inconsistent with other indications (no deduction will be made if gears are not led out)
XIII. Cooling device		
Fans or oil pumps should have rotational operating records	Data consulting	Deduct 0.1 point if the rotational operating records of the cooling device are not standardized; Deduct 0.5 point for each missing content of the rotational operating records, up to a maximum of 2 points
Tubular transformer coolers should be flushed at least 1 time per year and before heavy loads	On-site inspection and data consulting	Deduct 1.5 points if the flushing is not carried out in accordance with the regulations and there is no overhaul record; Deduct 0.5 point for each incomplete overhaul record. (Integrity is checked from 2015)
Substation monitoring background and dispatching monitor can monitor the operation of the main transformer cooling device and can correctly upload signals after the cooler is completely stopped	On-site inspection and data consulting	Deduct 1 point for any item found to be non-conforming
Each cooling device should be able to operate normally, and the cooling system fans and pumps should be put into operation at a balanced time to avoid excessive loss of a single group		Deduct 1 point for any item found to be non-conforming
XIV. Oil pump		
The submersible pumps installed at height should be equipped with rain covers	On-site inspection	Rain cover should be installed. The proper and secondary incoming cable should be covered by 50 mm, 45° downward, and can not be directly drenched by rainwater. Deduct 1.5 points for any item found to be non-conforming
There should be no abnormal vibrations or noises		Deduct 1.5 points for any item found to be non-conforming

Continued

Evaluation Item	Means of Inspection	Evaluation Principle
XV. Oil flow relays		
Correct indication, no water ingress or moisture, necessary rain protection measures, no leakage in the dial, no hand jitter	On-site inspection	Rain cover should be installed. The proper and secondary incoming cable should be covered by 50 mm, 45° downward, and can not be directly drenched by rainwater. Deduct 1.5 points for any item found to be non-conforming
Cooling unit control box		
The core of the control line cable is not exposed. If multi-core soft copper wire is used, insulating terminals shall be crimped	On-site inspection	Deduct 1.5 points for any item found to be non-conforming
The grounding of the terminal box body, secondary grounding inside the box, and the grounding connection wire between the box door and the body are well connected		Deduct 1.5 points if the grounding connection is not made; Deduct 0.5 point if the use of grounding wires is not standardized or the grounding parts are unreliable or rusted
The moisture driving device and heating device work normally, the temperature and humidity set value is set correctly, and the device is mobilized/demobilized according to the on-site operation regulations		Deduct 1.5 points for any item found to be non-conforming
The door sealing and internal plugging shall be good, without water intake, moisture, dust accumulation and foreign matters. If the door lamp used is a fluorescent lamp, a protective shield shall be installed		Deduct 1.5 points for any item found to be non-conforming
XVI. Gas relay		
Measure the insulation resistance of the secondary circuit of the gas relay according to the specified period	Data consulting	Deduct 3 points for failure to measure insulation resistance; Deduct 1 point for each failure to take measurements according to the period, up to a maximum of 3 points

Continued

Evaluation Item	Means of Inspection	Evaluation Principle
It should meet secondary circuit insulation resistance of more than or equal to 1 MΩ	Data consulting	Deduct 3 points if the test data does not meet the requirements and is not processed
Rainproof measures (excluding boxin and indoors)	On-site inspection	Rain cover should be installed. The proper and secondary incoming cable should be covered by 50 mm, 45°downward, and can not be directly drenched by rainwater. Deduct 1.5 points for any item found to be non-conforming
The observation windows should be open	On-site inspection	Deduct 1.5 points for any item found to be non-conforming
XVII. Pressure relief valve		
Measure the insulation resistance of the secondary circuit of the pressure relief valve according to the specified period	Data consulting	Deduct 3 points for each failure to measure insulation resistance according to the period
It should meet secondary circuit insulation resistance of more than or equal to 1 MΩ		Deduct 3 points if the test data does not meet the requirements and is not processed
The proper's pressure relief valve guide tube should not spray directly into the routine inspection channel, threatening the safety of O&M personnel, nor into the cable trench, busbar and other equipment	On-site inspection	Deduct 2 points for any item found to be non-conforming
XVIII. Oil flow speed relay		
Measure the insulation resistance of the secondary circuit of the oil flow relay according to the specified period	Data consulting	Deduct 3 points for failure to measure insulation resistance according to the specified period
It should meet secondary circuit insulation resistance of more than or equal to 1 MΩ		Deduct 3 points if the test data does not meet the requirements and is not processed
Rainproof measures (excluding boxin and indoors)	On-site inspection	Rain cover should be installed. The proper and secondary incoming cable should be covered by 50 mm, 45°downward, and can not be directly drenched by rainwater. Deduct 1.5 points for any item found to be non-conforming

Continued

Evaluation Item	Means of Inspection	Evaluation Principle
XIX. Thermometer		
Measure the insulation resistance of the secondary circuit of the thermometer according to the specified period	Data consulting	Deduct 3 points for each failure to measure insulation resistance according to the period
It should meet secondary circuit insulation resistance of more than or equal to 1 MΩ		Deduct 3 points if the test data does not meet the requirements and is not processed
Rainproof measures (excluding boxin and indoors)	On-site inspection	Rain cover should be installed for the thermometer and temperature probe. The proper and secondary incoming cable should be covered by 50 mm, 45°downward, and can not be directly drenched by rainwater. Deduct 1 point for any item found to be non-conforming
XX. Sudden pressure relays		
Measure the insulation resistance of the secondary circuit of the sudden pressure relay according to the specified period	Data consulting	Deduct 1.5 points for failure to measure insulation resistance according to the specified period
It should meet secondary circuit insulation resistance of more than or equal to 1 MΩ		Deduct 1.5 points if the test data does not meet the requirements and is not processed
Rainproof measures (excluding boxin and indoors)	On-site inspection	Rain cover shall be installed. The proper and secondary incoming cable shall be covered by 50 mm, 45° downward, and can not be directly drenched by rainwater. Deduct 1 point for any item found to be non-conforming
XXI. Secondary circuit		
There should be no waterlogged bends or over hanging of the secondary cable tube and a good job of temporary blocking and drainage hole opening should be done if so	On-site inspection	Deduct 1.5 points for any item found to be non-conforming
Secondary element markings should be clear and accurate		Deduct 1 point for any item found to be non-conforming

Continued

Evaluation Item	Means of Inspection	Evaluation Principle
XXII. On-line monitoring device for dissolved gas in transformer oil		
It should be connected to the power transmission and transformation status monitoring system	On-site inspection and data consulting (No points will be deducted for no gas-in-oil component device)	Deduct 1 point for any item found to be non-conforming
It should operate normally (no oil leakage, undervoltage and insufficient air) and the data should be uploaded accurately		Deduct 1 point for any item found to be non-conforming
The sampling period should meet the requirements. Class I substation: 4 hours; Class II substations: 12 hours; Class III substation: 24 hours		Deduct 1.5 point if the sampling cycle is non-conforming
Operation and maintenance records are available		Deduct 1 point for any item found to be non-conforming
The connection piping between the transformer propers should be well protected		Deduct 1 point for any item found to be non-conforming
XXIII. Grounding		
Major parts and components should be short-circuited to ground	On-site inspection	Bell or barrel, oil storage cabinet, bushing, enclosure, load switch and terminal box should be short-circuited to ground and deduct 1 point for any item found to be non-conforming
Proper grounding		There should be two horizontal grounding bodies leading to different locations at different positions, and deduct 1 point if they do not meet the requirements
Neutral point grounding		There should be two grounded down conductors connected to different sides of the main grid of the ground grid. If it does not meet the requirements, deduct 1 point. The neutral point overvoltage protection of the 110 to 220 kV non-grounded transformer should be protected by rod clearance, the clearance distance and lightning arrester parameters should be checked, and if it does not meet the requirements, deduct 1 point

Continued

Evaluation Item	Means of Inspection	Evaluation Principle
The conductive cross section of the grounding device is free of corrosion; The conductor contact is good	On-site inspection	Deduct 0.5 point for any item found to be non-conforming
The guide rail foundation proper should be well fixed		Deduct 1 point for any item found to be non-conforming

Points for attention in work:

(1) Pay attention to distinguish the current tap changer type and manufacturer, the mechanism of different manufacturers. The internal component layout and electrical circuit will be different, and the working plan and defect elimination idea formulated shall be changed according to the actual situation (See Fig. 2-59, 2-60, 2-61).

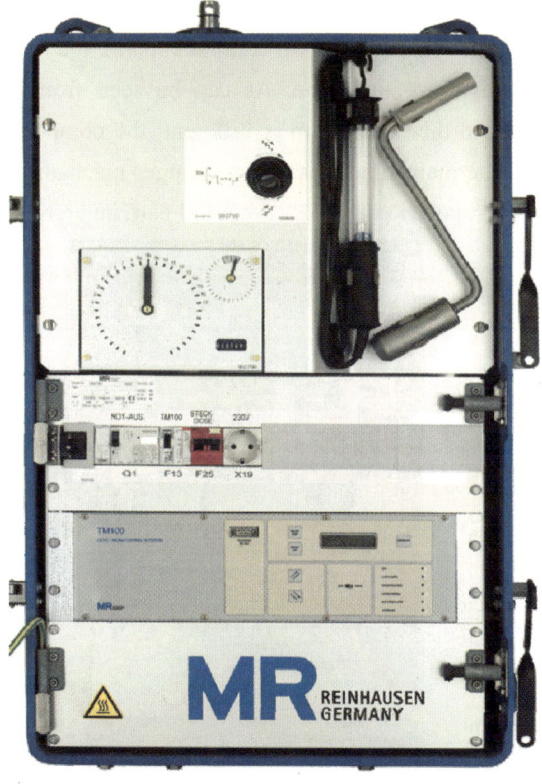

Fig. 2-59 MR Company MA7 Mechanism (with Panel)

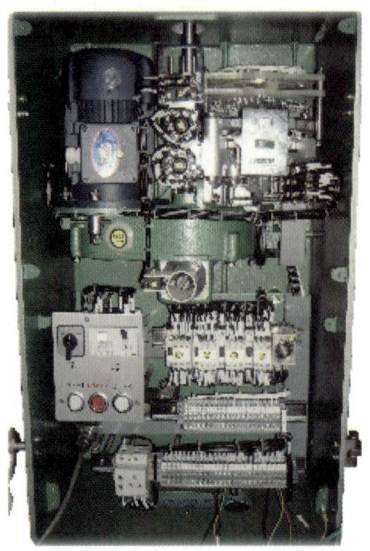 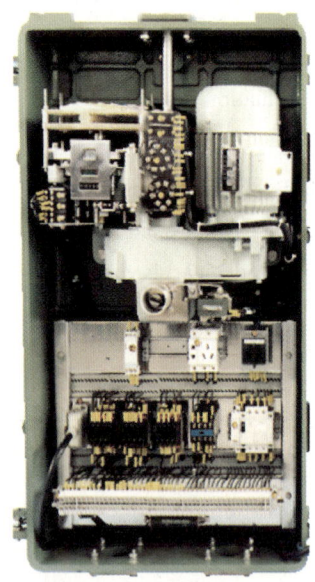

Fig. 2-60　MA7 Mechanism of Huaming Company (Panel Removed)

Fig. 2-61　DCJ10 Mechanism of Changzheng Company (Panel Removed)

(2) Pay attention to the distribution of components in the mechanism. Most of the components of the mechanism are the same. As can be seen from Fig. 2-62, although the arrangement of components in the mechanism is different, the component types are almost the same. Generally speaking, the main functions of tap changer mechanism (whether M type or V type) shall include three types: power part, process control part, and remote control wiring part.

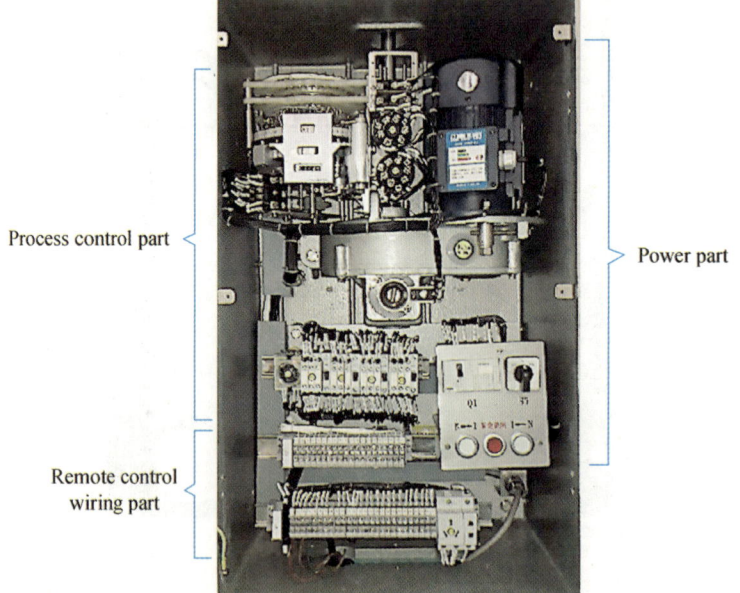

Fig. 2-62　Function Distribution Modules within the Mechanism

(3) The complete gear confirmation shall be that the remote control display gear, the window display gear in the mechanism, and the sight glass display gear of the switch top are consistent. Strictly speaking, the gear of the "Remote control" operation and display end is only the "Signal" performance of the feedback of various contactors and travel switches during the "Local" operation. If safety conditions permit, as long as the display of the gear position on the mechanism and the gear position of the switch top are consistent, it can be considered that the gear matching is correct, as shown in Fig. 2-63. Incorrect remote control display is often due to electrical signal or wiring issues.

(a) Top display gear

(b) Mechanism display gear

(c) Remote control display gear

Fig. 2-63　Gear Display (All Gears Are Gear 7)

(4) The heating resistor is generally on the inside of the mechanism, and sometimes on the inner wall of the mechanism, as shown in Fig. 2-64.

Fig. 2-64　Location of Heating Resistor

(5) The position of applying lubricating grease to the gear is as shown in Fig. 2-65.

(a) Applying position

(b) Applying state

Fig. 2-65　Location of Gear Grease Application

(6) When checking the electrical connections, it is not only necessary to check the firmness of the wiring, but also to pay attention to whether the supporting gaskets or spring pads are complete, as shown in Fig. 2-66 and Fig. 2-67.

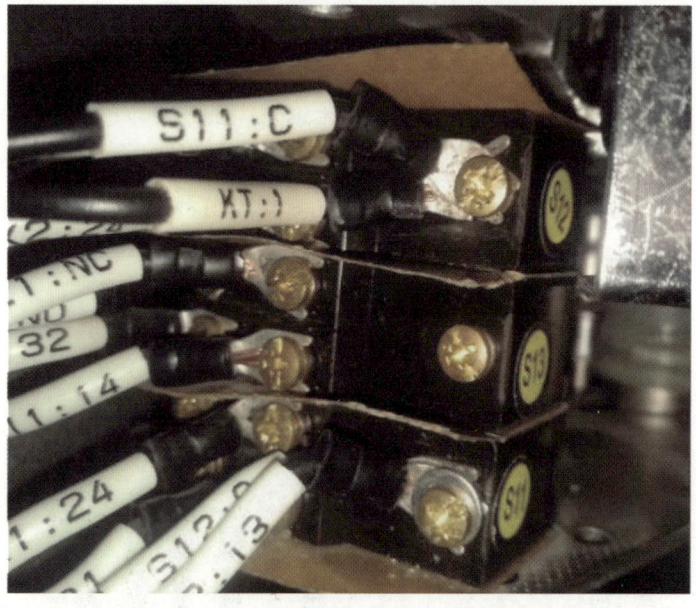

Fig. 2-66　All Terminal Gaskets and Spring Washers of the Travel Switch Are Missing

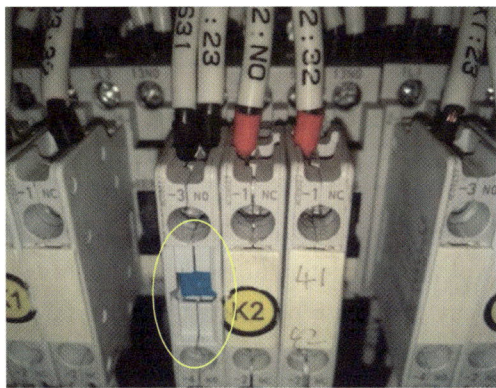

Fig. 2-67 Contactor Manual Bounce Contact Terminal Cover Is Missing

(7) The method of connection check (gear adjustment) is not only used to verify the function of the normal gear, but also suitable for the adjustment of the sliding gear (gear subject to deviation). The specific methods are as follows:

1) In the case of good mechanical coordination: in any position (V-type switch recommended to be in the set gear) at the normal mechanism display gear (that is, the red mark line is in the middle of the window), manually upshift the gear until the next normal display gear. The number of rotations of the operating crank during this period is recorded as X, and the gear is downshifted back to the starting gear. The number of rotations during this period is recorded as Y, and the intact mechanical fit is reflected as $|X-Y|\approx 0$. But actually, because there are many force transmission parts, like gear, link, transmission box and transmission shaft, driven by mechanism transmission, it is difficult to realize $|X-Y|\approx 0$, so in general work, if $|X-Y|\leqslant 1$ is realized, that is, the difference in the number of turns is not greater than 1 turn after shift up and down once each, it is considered to be a good fit.

2) In the case of deviation in mechanical coordination: count X and Y according to "1)", remove the drive link, use the operating handle to idle in the direction of large values of X and Y, and the number of turns is $|X-Y|\div 2=N$. Connect the drive link again, verify according to the step of "①", and repeat the step of "②" until $|X-Y|\leqslant 1$.

3) If you have a good understanding of the switching state of the tap changer (see "Related theoretical knowledge"), there is a simpler way to do it. Because the switching of the tap changer must wait until switching of the switch contact terminal is completed before the power of the mechanism motor can be cut off, otherwise, the contact terminal will be suspended and discharged until it is burned out, and the contact terminal switching process of each gear of the tap changer is the same, the main difference in the travel of each gear is the period from "switching of contact terminal completed" to "mechanism motor power off", this period of time is referred to as the 'inertia travel' (idle travel), that is, the number of turns of X or Y is composed of "switch switching process + inertia travel", $|X-Y|=|$ "switch switching

process (up)+inertia travel (up)" – "switch switching process (down)+inertia travel (down)"|. According to the contact terminal switching process of each gear described above, the switching process is the same, so "switch switching process (up)" = "switch switching process (down)", then $|X-Y|$ = the difference between the inertia travel of the up and down gear.

The inertia travel starts counting when the contact terminal switching is completed, that is, after hearing the sound of the completion of the switching, or feeling a slight vibration on the handle (which is the energy release of the fast mechanism spring). According to this method, the inertia travel count of the up and down gears is greatly reduced.

任务三 变压器呼吸器（吸湿器）小修

一、工作任务

根据任务要求，对 35 kV 及以上变压器呼吸器（吸湿器）进行硅胶粒更换。

二、引用标准

（1）《国家电网公司电力安全工作规程（变电部分）》（Q/G DW1799.1—2013）。

（2）国家电网公司生产技能人员职业能力培训专用教材——《变压器检修》。

（3）《国家电网公司变电检修管理规定（试行）第 1 分册油浸式变压器（电抗器）检修细则》。

（4）《国家电网公司变电运维管理规定（试行）第 1 分册油浸式变压器（电抗器）运维细则》。

（5）《电力变压器检修导则》（DL/T 573—2010）。

（6）《变压器分接开关运行维修导则》（DL/T 574—2010）。

三、现场气候及工作要求

（1）新装吸附剂应干燥，其颗粒直径 4～7 mm。

（2）回收吸附剂：①置入烘箱干燥，干燥温度从 120℃升至 160℃，时间 5h。②吸附剂变成蓝色方可使用。

四、工作准备

（一）危险点及预控措施

（1）危险点——触电伤害。预控措施如下：

① 主变未停电时的机构维护检查不能爬上主变观察档位，只能从远控端观察档位；

② 确定临时电源位置，电缆放线尽量从工作区域以外绕行；

③ 操作前用万用表验证机构电源侧已断开；

④ 班组若自备开关箱注意检查接线和接地，确认空开处于断开位置，通电时，先合上通临时电源空开，再合上开关箱空开，最后合上机构箱空开；

⑤ 所有工作用电缆接线需检查绝缘完好状况，若有破裂，用黄色绝缘防水胶带多层包扎，并在下面垫干燥木板。

（2）危险点——防止滑倒摔伤。预控措施如下：

① 工作人员应正确穿戴防滑鞋；

② 地面若有油污，进出工作现场、上下梯子需要清洁鞋底；

③ 为防止机构零件掉落入池内无法寻找，在机构箱下的防火池地面上根据操作范围适量铺设白布或蛇纹布，尽量不用油布，零件掉落在油布上可能被弹开掉入防火池，且在沾油的油布上行走更容易滑倒；

④ 在防火池内架设梯子或绝缘高凳，必须有专人扶持。

（二）工器具及材料选择

M 型有载分接开关机构箱检修工器具及材料见表 2-10。

表 2-10　M 型有载分接开关机构箱检修工作的工器具及耗材表

类别	名称	规格型号	数量	备注
专用工具	安全帽		3 顶	
	绝缘手套		1 副	
	兆欧表（或电子式）	500 V 或 1 000 V	1 套	
个人工具	平口改刀		1 把	回路端子
	十字改刀		1 把	拆面板
	呆扳手	8-10,10-12,17-19	各 2 把	拆除连接件
	活络扳手	300 mm	1 把	辅助活络扳手使用
	工具盘		1 张	
耗材	硅胶		1 盒	
	密封圈		1 套	
	变压器油		1 桶	
其他	无水乙醇		1 瓶	
	白布		若干 m²	

五、工作程序

工作程序见表 2-11。

表 2-11 工作程序

序号	作业内容	作业程序
1	吸湿器拆卸	1. 拆卸吸湿器 2. 放掉油杯里面油 3. 会将底座和玻璃管取下
2	吸湿器检查	油杯、底座和玻璃管检查
3	连接管检查	上法兰与连接管畅通，无锈蚀，无堵塞现象
4	清理	清理部件
5	硅胶更换	1. 更换变色的硅胶 2. 更换胶垫 3. 对硅胶容量进行检查(留有 1/5-1/6 空隙)
6	油杯换油	1. 注入合格变压器油 2. 油杯内装变压器油量合格
7	吸湿器安装	1. 吸湿器安装时螺栓受力均匀 2. 油杯安装到位后松半圈（仅旧结构油杯需要） 3. 胶垫放置位置合格 4. 胶垫不得破损，破损的需要更换 5. 检查呼吸器是否呼吸顺畅

吸湿器（见图 2-68）是一个圆形的容器，上端通过联管接到变压器的储油柜上，下端有孔与大气相通，其主体为玻璃管，内部盛有变色硅胶(或活性氧化铝)作为干燥剂。其下部带有油杯(盛油器)，作为空气进口处的过滤装置。当变压器由于负载或环境温度的变化而使变压器油体积发生胀缩时，储油柜内的气体通过吸湿器来吸气和排气。

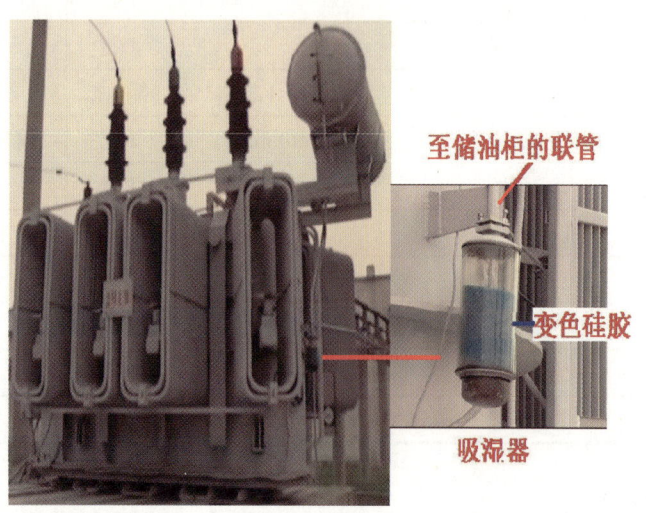

图 2-68 吸湿器

Task 3   Minor Repair of Transformer Breather (Moisture Absorber)

3.1   Work Tasks

According to the task requirements, replace silicone particles of breathers (moisture absorbers) of 35 kV and above transformers.

3.2   References

(1) *Electric Power Safety Working Regulations (Power Transformation) of State Grid Corporation of China* (Q/GDW 1799.1-2013).

(2) Specialized teaching materials for the training of vocational competence of production skill personnel of the state grid of China: *Transformer Maintenance*.

(3) *Substation Maintenance Management Regulations of State Grid Corporation of China (Trial) - Volume 1: Detailed Rules for Maintenance of Oil Immersed Transformers (Reactors)*.

(4) *Regulations of State Grid Corporation of China on Management of Substation Operation and Maintenance (Trial) - Volume 1: Detailed Rules for Operation and Maintenance of Oil Immersed Transformers (Reactors)*.

(5) *Maintenance Guide for Power Transformer* (DL/T 573-2010).

(6) *Guide for the Operation and Maintenance of Tap Changers in the Power Transformer* (DL/T 574-2010).

3.3   Site Environment and Work Requirements

(1) The newly installed adsorbent shall be dry and have a particle diameter of 4~7 mm.

(2) Recovery of adsorbent: Place it in an oven for drying, and increase the drying temperature from 120°C to 160°C for 5h. The absorbent cannot be used until it turns blue.

3.4   Preparation for Work

1. Danger points and preventive and control measures

(1) Danger point - electric shock injury. The preventive and control measures are as follows:

① When the main transformer is not de-energized, the operator shall not climb up to the main transformer to observe the gears for mechanism maintenance and inspection, but can only observe the gears from the remote control end;

② Determine the location of the temporary power supply and try to pay off cables beyond the work area.

③ Use a multimeter to verify that the power supply side of the mechanism is disconnected before operation;

④ If the team is provided with its own switch box, the team members shall check the wring and grounding, and confirm that the air switch is in the "OFF" position. To turn on the power, first close the air switch for temporary power supply, and then close the air switch for switch box, and finally close the air switch for mechanism box;

⑤ For all the cable connections used for operation, check whether the insulation is in good condition. If there is any fracture, wrap it with yellow insulating and water-proof tape, and place a dry board under it.

(2) Danger point - slipping and falling. The preventive and control measures are as follows:

① The staff shall wear non-slip shoes properly;

② If there is any oil dirt on the floor, clean shoe soles before entering and leaving the workplace or climbing up and down the ladder;

③ To prevent parts of the mechanism from falling into the tank, lay appropriate white cloth or serpentine cloth on the fireproof pool under the mechanism box according to the range of operation. Try not to use tarpaulin since parts may bounce and fall into the fireproof pool after falling on the tarpaulin. Besides, people who walk on the tarpaulin with oil may slip and fall more easily;

④ When setting up a ladder or insulated high stool in the fireproof pool, the ladder or insulated high stool must be supported by a specially-assigned person.

2. Selection of tools, instruments and materials

For tools, instruments and materials used for maintenance of M-type on-load tap changer mechanism, see Table 2-10.

Table 2-10　Tools, Instruments and Maintenance Used for Maintenance of M-Type On-Load Tap Changer Mechanism Box

Type	Name	Specification and model	Quantity	Remarks
Special tools	Safety helmet		3 pieces	
	Insulating gloves		1 pair	
	Megger (or electronic type)	500V or 1,000V	1 set	
Personal tools	Slotted screwdriver		1 set	Circuit terminal
	Cross screwdriver		1 set	For panel removal
	Open spanner	8-10,10-12,17-19	2 pieces for each	For removal of connecting parts
	Adjustable spanner	300mm	1 set	As an auxiliary tool for adjustable spanner
	Tool tray		1 piece	

Continued

Type	Name	Specification and model	Quantity	Remarks
Consumables	Silica gel		1 box	
	Seal ring		1 set	
	Transformer oil		1 barrel	
Others	Absolute ethyl alcohol		1 bottle	
	White cloth		Several m^2	

3.5 Operation Procedures

Operation procedures see Table 2-11.

Table 2-11 Operation Procedures

S/N	Scope of work	Operation procedures
1	Remove the moisture absorber	1. Remove the moisture absorber 2. Drain the oil from the oil cup 3. Can remove the base and glass tube
2	Check the moisture absorber	Check the oil cup, base, and glass tube
3	Check the connecting tube	The upper flange and connecting tube are unobstructed, without rust or blockage
4	Cleaning	Clean components
5	Replace silica gel	1. Replace the discolored silica gel 2. Replace the rubber pad 3. Check the capacity of silica gel (leaving 1/5-1/6 gap)
6	Change oil in oil cup	1. Inject qualified transformer oil 2. The oil quantity of the transformer in the oil cup is qualified
7	Installation of moisture absorber	1. The bolts are evenly stressed during the installation of the moisture absorber 2. Loosen the oil cup by half a circle after installation (only required for old-structure oil cup) 3. The placement position of the rubber pad is acceptable 4. The rubber pad must not be damaged, and damaged ones need to be replaced 5. Check whether the breather is breathing smoothly

The moisture absorber (see Fig. 3-68) is a round container, the upper end is connected to the oil conservator of the transformer through the connecting pipe, and the lower end has a hole that is connected to the atmosphere. Its main body is a glass tube, which contains color-changing silica gel (or activated alumina) as a drying agent. There is an oil cup (oil holder) in its lower part, which serves as a filter unit at the air inlet. When the volume of transformer oil expands or contracts due to changes in the load or ambient temperature, the gas in the oil conservator is aspirated and vented through the moisture absorber.

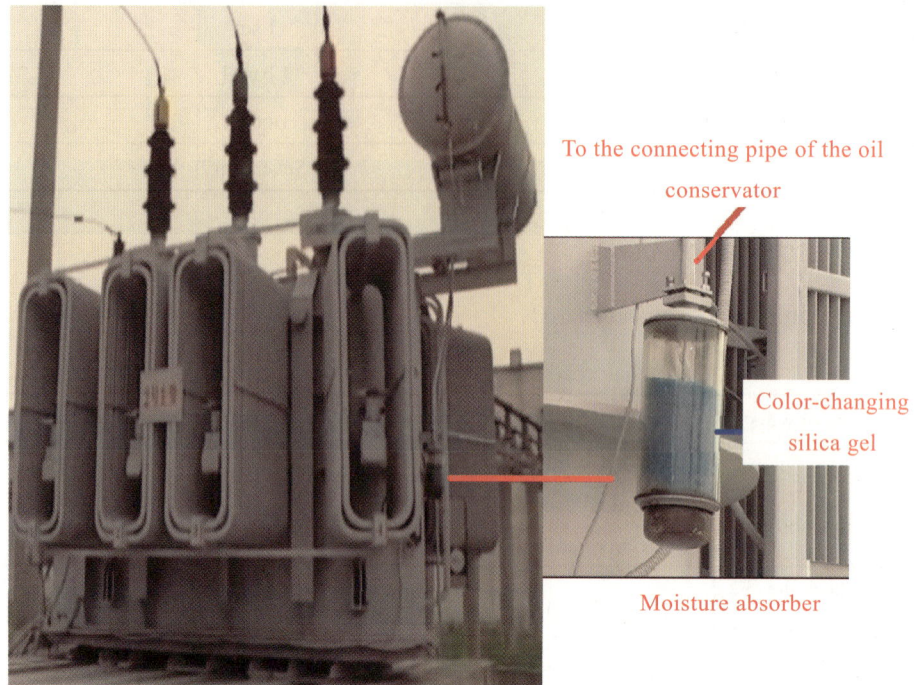

Fig. 3-68 The Moisture Absorber

项目三　互感器检修

模块一　互感器概述

一、互感器的分类

互感器按性质主要分为电压互感器和电流互感器两大类。也有把电压互感器和电流互感器合并形成一体的互感器，称为组合式互感器。

二、互感器的作用

互感器在电力系统广泛使用，他是一种利用电磁原理进行电压、电流变换的变压器类设备（光电互感器除外）。互感器与测量仪表和计量装置配合，可以测量一次系统的电压、电流和电能；与继电保护和自动装置配合，可以对电网各种故障进行电气保护以实现自动控制。其作用归纳为：

（1）将一次系统的电压或电流信息准确地传递到二次设备。

（2）将一次系统的高电压或大电流变换为二次侧的低电压或小电流，使二次设备装置标准化、小型化，并降低了对二次设备的绝缘要求。

（3）由于互感器一、二次之间有足够的绝缘强度，能使二次设备和工作人员与一次系统设备在电方面很好地隔离，从而保证了二次设备和工作人员的人身安全。

Program 3　Transformer Maintenance

Module 1　Transformer Overview

3.1.1　Classifications of Transformers

By properties, transformers are mainly divided into two categories: voltage transformer and current transformer. There is also a kind of transformer which combines the voltage transformer and the current transformer, which is called the combined transformer.

3.1.2　Functions of Transformers

Transformer is widely used in power system. It is a kind of transformer equipment that uses electromagnetic principle to transform voltage and current (except photoelectric transformer). The transformer, in conjunction with measuring instruments and measuring devices, can measure the voltage, current, and electrical energy o f the primary system; The transformer, in conjunction with relay protection and automatic devices, can provide electrical protection for various faults in the power grid to achieve automatic control. Its functions can be summarized as follows:

(1) Accurately transmit the voltage or current information of the primary system to the secondary equipment.

(2) Transform the high voltage or high current of the primary system into the low voltage or low current of the secondary side, standardize and miniaturize the secondary equipment, and reduce the insulation requirements for the secondary equipment.

(3) Due to the sufficient insulation strength between the primary and secondary sides of the transformer, it can effectively isolate the secondary equipment and staff from the primary system equipment in terms of electricity, thereby ensuring safety of the secondary equipment and personal safety of staff.

模块二　电压互感器

一、电压互感器的主要技术数据

（一）电压互感器的分类

目前，电压互感器的分类按不同情况划分如下。

1. 按用途分类

测量用电压互感器和保护用电压互感器，这两种电压互感器，又可分为单相电压互感器和三相电压互感器。

2. 按安装地点分类

户内型电压互感器和户外型电压互感器。

3. 按电压变换原理分类

（1）电容式电压互感器，以电容分压来变换电压。
（2）光电式电压互感器，以光电元件来变换电压。
（3）电磁式电压互感器，以电磁感应来变换电压。
电磁式电压互感器是本章节重点介绍的内容，以后凡是未加特殊说明的电压互感器，均指电磁式电压互感器。

4. 按结构分类

（1）单级式电压互感器，一次绕组和二次绕组均绕在同一个铁心柱上。
（2）串级式电压互感器，一次绕组分成匝数相同的几段，各段串联起来，一端子连接高压电路，另一端子接地。

（二）电压互感器的型号规定

目前，国产电压互感器型号编排方法如图 3-1 所示。

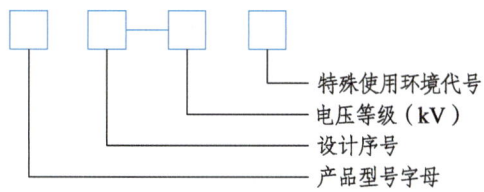

图 3-1　国产电压互感器型号编排方法

电压互感器型号中的字母，都用汉语拼音字母表示，字母排列顺序及其对应符号含义如表 3-1 所示。

电压互感器在特殊使用环境的代号，主要有以下几种：CY——船舶用；GY——高原地区用；W——污秽地区用；AT——干热带地区用；TH——湿热带地区用。

表 3-1　电压互解器型号字母的含义及排列顺序

序号	类别	含义	代表字母
1	名称	电压互感器	J
2	相数	单相	D
		三相	S
3	绕组外的绝缘介质	变压器油	不标注
		空气（干式的）	G
		浇注成固体形结构特征	Z
		气体	Q
4	结构特征	带备用电压绕组	X
		三柱芯带补偿绕组	B
		五柱芯每相三绕组	W
		串级式带备用电压绕组	C

电压互感器的主要参数如下。

1. 绕组的额定电压

额定一次电压是指可以长期加在一次绕组上的电压，并在此基准下确定其各项性能；根据其接入电路的情况，可以是线电压，也可以是相电压。其值应与我国电力系统规定的"额定电压"系列相一致。

额定二次电压，我国规定接在三相系统中相与相之间的单相电压互感器为 100 V，对于接在三相系统相与地间的单相电压互感器，为 $100/\sqrt{3}$ V。

2. 额定电压变比

额定电压变比为额定一次电压与额定二次电压之比，一般用不约分的分数形式表示为

$$K_U = \frac{U_{1e}}{U_{2e}}$$

3. 额定二次负载

电压互感器的额定二次负载，为确定准确度等级所依据的二次负载导纳（或阻抗）值。额定输出容量为在二次回路接有规定功率因数的额定负载，并在额定电压下所输出的容量，通常用视在功率（单位 V·A）表示。

实际测试中，电压互感器的二次负载常以测出的导纳表示，负载导纳与输出容量的关系为

$$S = U_2^2 Y$$

由于 U_2 的额定值为 100 V，故常可用 $S = Y \times 10^4$ 来计算。

4. 准确度等级

由于电压互感器存在一定的误差，因此根据电压互感器允许误差划分互感器的准确度等级。国产电压互感器的准确度等级有 0.2 级、0.5 级、1 级、3 级。

以上电压互感器，0.2 级主要用于试验时进行精密测量、计量，或者作为标准用来检验低等级的互感器，也可以与标准仪表配合，用来检验仪表；0.5 级一般用于发配电设备的测量；1 级和 3 级主要用于非精密测量、监控和保护制造厂在铭牌上标明准确度等级时，必须同时标明确定该准确度等级的二次输出容量。

5. 极性标志

为了保证测量及校验工作的接线正确，电压互感器一次及二次绕组的端子应标明极性标志。电压互感器一次绕组接线端子用大写字母 A、B、C、N 表示，二次绕组接线端子用小写字母 a、b、c、n 表示。

二、工作原理和误差特性

（一）工作原理

电压互感器的工作原理、结构和接线方式与电力变压器相似，同样是由相互绝缘的一次、二次绕组绕在公共的闭合铁心上组成的，如图 3-2 所示。其主要区别是二者容量不同，且电压互感器是在接近空载的状态下工作的。

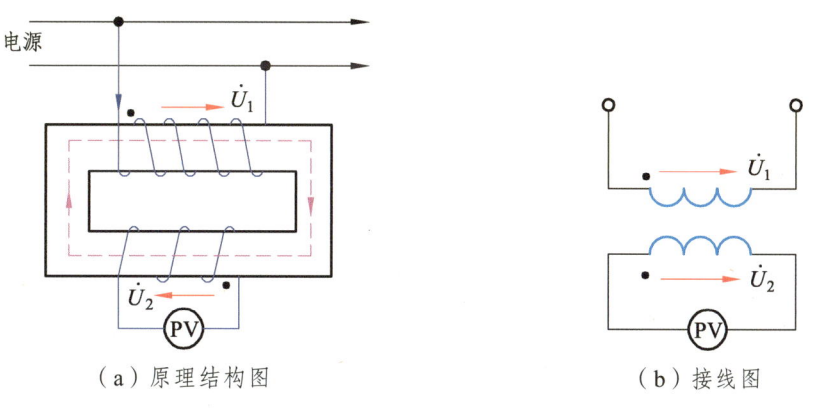

（a）原理结构图　　　　　　　　（b）接线图

图 3-2　电压互感器的原理结构图和接线图

电压互感器将高电压变为低电压供电给仪表，所以它的一次匝数 N_1 多，二次匝数 N_2 少。一次绕组与被测电压并联，二次绕组与各种测量仪表或继电器的电压线圈相并联。电压互感器的二次侧应装设熔断器，以保护自身不因二次绕组短路而损坏；在有可能的情况下，一次侧也应装设熔断器，以保护高压电网不因互感器一次绕组或引线故障危及一次系统安全。电压互感器在电气图中文字符号用 TV/PT 表示。

当一次绕组加上电压时，铁心内有交变主磁通通过，一、二次绕组分别有感应电动

势 $\dot{E}_1$ 和 $\dot{E}_2$。将电压互感器二次绕组阴抗折算到一次侧后，可以得到分别如图 3-3 和图 3-4 所示的 T 形等值电路图和相量图。

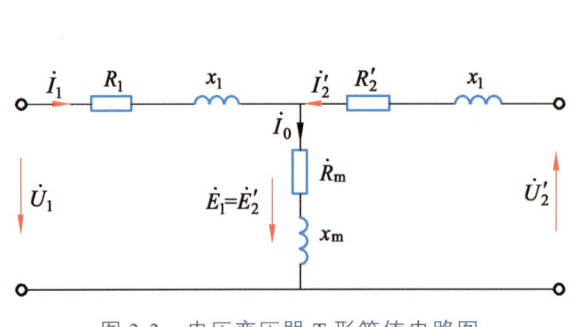

图 3-3　电压变压器 T 形等值电路图

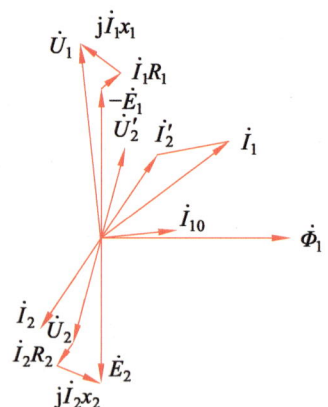

图 3-4　电压变压器相量图

从等值电路图中得到

$$\dot{U}_1 = \dot{I}_1(R_1 + jX_1) - \dot{E}_1$$

$$\dot{U}_2' = \dot{E}_2' - \dot{I}_2'(R_2' + jX_2')$$

式中　R_1、X_1——一次绕组的电阻和阻抗；

R_2'、X_2'——二次绕组折算到一次侧的电阻和阻抗。

若忽略励磁电流和负载电流在一、二次绕组中产生的压降得到

$$K_U = \frac{U_1}{U_2} = \frac{E_1}{E_2} = \frac{N_1}{N_2}$$

这就是理想电压互感器的电压变比，称为额定变比，即理想电压互感器一次绕组电压 U_1 与二次绕组电压 U_2 的比值是个常数，等于一次绕组和二次绕组的匝数比。

实际上，电压互感器是有铁损和铜损的，绕组中有阻抗压降。从相量图 3-4 中看出，二次电压旋转 180°以后（$-\dot{U}_2'$）与一次电压大小不等，且有相位差，就是说电压互感器存在着比差和角差。

比差用 f_U 表示，它等于

$$f_U = \frac{U_2' - U_1}{U_1} \times 100\% = \frac{\frac{N_1}{N_2}U_2 - U_1}{U_1} \times 100\% = \frac{K_U - K_U'}{K_U'} \times 100\%$$

式中　U_1——实际一次电压有效值；

U_2——实际二次电压有效值；

K_U'——实际电压互感器变比，$K_U' = \frac{U_1}{U_2}$；

K_U——额定电压互感器变比，$K_U = \dfrac{U_{1e}}{U_{2e}} = \dfrac{N_1}{N_2}$。

相角差简称角差，是指一次电压与旋转 180° 后二次电压相量间的相位差，用 δ_U 表示，单位为 "′"（分）。当旋转后的二次电压超前于一次电压相量时，角差为正值；反之，角差为负值。

（二）误差特性

1. 电压互感器的负载特性

电压互感器二次负载电流的大小和性质，对比差和角差均有影响。图 3-5 为一次电压 $\dot{U}_1$ 不变时，比差、角差随二次电流的变化关系。图中空载（$I_2 = 0$）时，比差为负，角差为正。空载时的比差和角差是由互感器的结构和空载电流决定的，空载电流越大，则比差向负的方向增加，角差向正的方向增加。

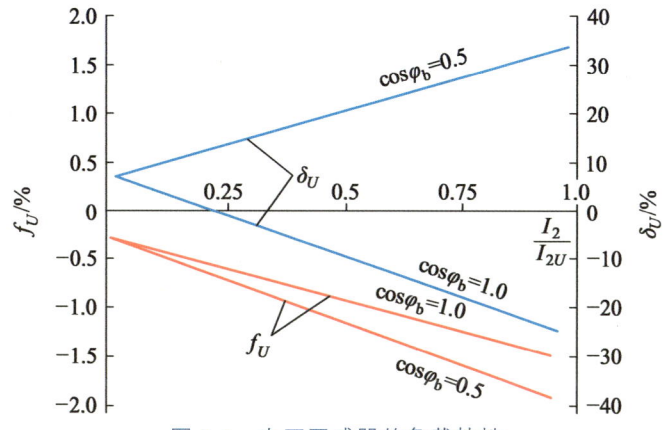

图 3-5 电压互感器的负载特性

随着负载电流的增加，比差要在空载的基础上继续向负的方向增加，且功率因数越低，向负的方向增加得越多。而角差在功率因数较低时，随负载电流的增加总是向正的方向增加的；当功率因数较高时，则先由正值变为零再向负的方向增加。

由于二次负载对比差、角差产生影响，是二次电流在绕组中产生的电压降所致。因此，限制绕组导线的电流密度，减小绕组的漏磁以降低漏抗，是提高电压互感器准确度的有效措施。

2. 电压互感器的电压特性

电压特性就是电压互感器的比差和角差与一次电压的关系，其变化趋势如图 3-6 所示。一次电压对比差和角差的影响，可由铁心的磁化曲线得到解释。如图 3-7 所示，由于铁心的导磁率 μ 和损耗角 α 均为非线性，所以，对同样的磁场强度增量 ΔH，对应磁化曲线不同部分的磁感应强度增量 ΔB 是各不相同的。当一次电压较低时，由于导磁率 μ 较低，所需励磁电流 I_0 较大，故空载时的比差 f_0 和角差 δ_0 均较大。随着电压的增加，导磁率增加，铁心工作在磁化曲线的平直部分，所以，比差和角差开始减小并逐渐趋于平稳。可见，应使电压互感器一次侧工作于额定电压下。

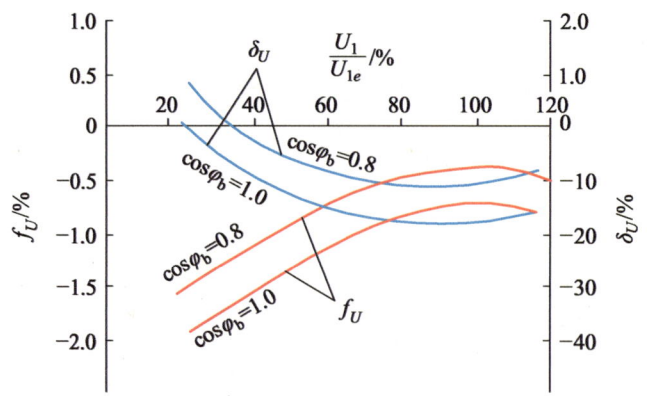

图 3-6　电压互感器的电压特性

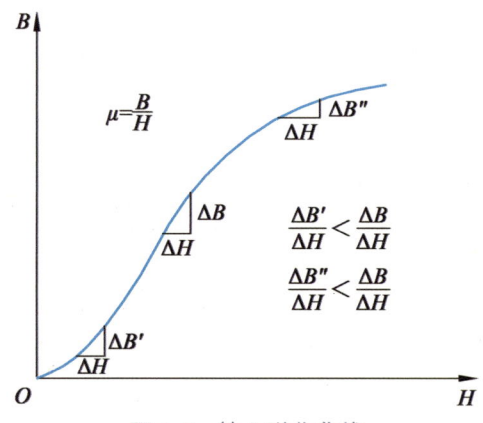

图 3-7　铁心磁化曲线

（三）电压互感器误差的补偿方法

如上所述，空载电流和负载电流在电压互感器一、二次绕组阻抗上，特别是在电阻上产生的电压降，是引起比差和角差的主要原因。I_0 越大，则空载误差（也称固有误差）越大。内部参数（R_1、R_2、X_1、X_2）越大，负载电流增加时会引起较大的电压降，使误差增加。此外，R_1、R_2 与 X_1、X_2 比值的变化，对角差也要产生影响。因此，为提高电压互感器的准确度，从设计方面可采取以下措施。

（1）铁心选用导磁率较高的材料，如单晶冷轧硅钢片，含镍 50% 的坡莫合金等。
（2）适当增加绕组导线截面积，以减小电阻。
（3）绕组装配尽量紧凑、均匀，以减小漏抗。

以上措施必然会增加制造成本。为此，还可以采取以下几种误差补偿方法。

1. 匝数补偿法

由于电压互感器一、二次电压与绕组的匝数成正比，故适当地增加或减少绕组的匝数，电压也将相应地增加或减小。例如，二次绕组适当增加，则二次电压增加，当一次电压不变时，则实际变比 K_U 有所下降。若适当减少二次绕组匝数，则比差要向负的方向变化，使比差特性变坏。改变一次绕组匝数对比差的影响效果与改变二次绕组匝数恰好相反。

为使调整幅度不致过大，可将二次绕组用双股或多股导线绕制。当用双股或 n 股导线增加（或减少）匝数，便可收到 1/2 或 $1/n$ 匝的补偿效果。在实际调整中，为获得所需的比差特性，一般是在 50%额定负载时，把比差调到零。这样，当负载较小时比差为正，当负载较大时比差为负。因此，比差特性可得到改善。

应指出，匝数补偿法只能改善比差特性，而不能改善角差特性。

2. 附加绕组补偿法

如图 3-8（a）所示，将一个附加绕组（又称补偿绕组）N_K 经电阻 R 并联于二次绕组 N_2 上。令 N_K 比 N_2 少一匝或几匝，于是它们产生的感应电动势 $\dot{E}_K < \dot{E}_2$，但两个电动势的相位是相同的。二者差电动势为 $\Delta \dot{E}$，在 $\Delta \dot{E}$ 的作用下产生一平衡电流 $\dot{I}_P$，如图 3-8（b）所示，$\dot{I}_P$ 的大小和相位由电阻 R 和两个绕组 N_K 和 N_2 的内阻抗决定。因此，改变电阻 R 和 N_K、N_2 的匝数，便可改变 $\dot{I}_P$ 的大小和相位，从而可改变二次电压 $\dot{U}_2$ 的大小和相位，于是比差 f_U 和角差 δ_U 随之改变。只要适当选择 N_K 和 N_2 的匝数及 R 的数值，便可将 f_U 和 δ_U 补偿到满意的程度。

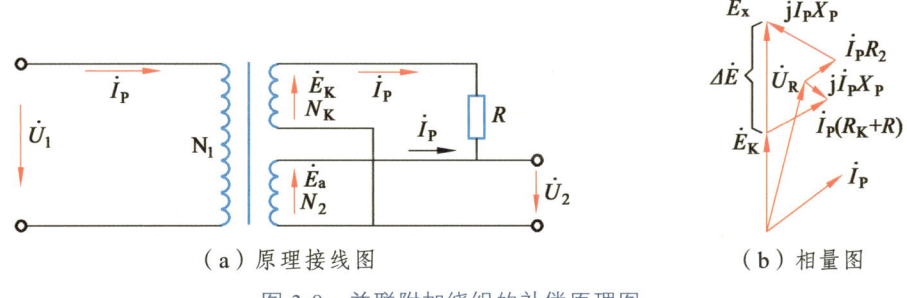

（a）原理接线图　　　　　（b）相量图

图 3-8　并联附加绕组的补偿原理图

3. 二次绕组并联外阻抗补偿法

一般是采用并联电容器进行补偿，如图 3-9 所示。并联电容 C 以后，二次绕组中要产生超前其二次电压 90°的电容电流 $\dot{I}_C$。根据磁动势平衡原理，在一次绕组中必然增加一个电流 $-\dot{I}'_C$ 以便与 $\dot{I}'_C$ 相平衡，$\dot{I}_C$ 和 $-\dot{I}'_C$ 分别流过一、二次绕组时要产生阻抗电压降 $\dot{I}_C Z_2$ 和 $-\dot{I}_C Z_1$，因此，要引起附加误差。如果将 $\dot{I}_C Z_2$ 折算到一次侧变为 $\dot{I}'_C Z'_2$，并以复数误差表示电容电流引起的附加误差，则有：

$$\Delta \dot{\gamma}_U = -\frac{\dot{I}'_C (Z_1 + Z'_2)}{\dot{U}_1}$$

如图 3-9（b）所示，$\Delta \dot{\gamma}_U$ 的水平分量为附加代差 $\Delta \dot{f}_U$，它对 $\dot{f}_U$ 有补偿作用；$\Delta \dot{\gamma}_U$ 的垂直分量为附加角差 $\Delta \dot{\delta}_U$，它对 $\dot{\delta}_U$ 无补偿作用。补偿的数值与一、二次绕组的内阻抗

及电容电流的大小成正比而与二次负载无关,故并联电容器补偿法实际上只相当补偿了空载误差。

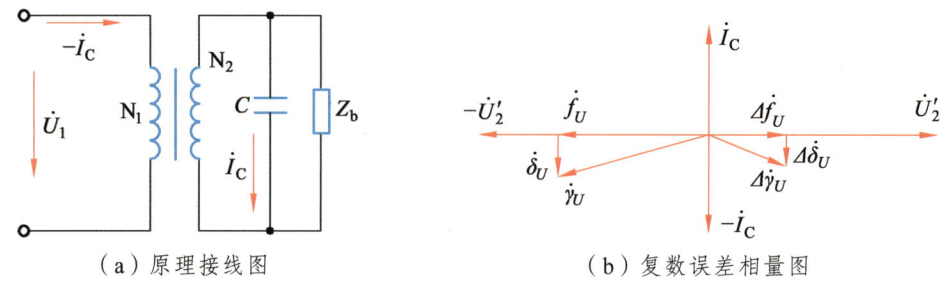

(a)原理接线图　　　　　　　　(b)复数误差相量图

图 3-9　并联外加阻抗的补偿原理

三、电压互感器的正确使用

1. 电压互感器的选择

(1)额定电压的选择。电压互感器的额定电压是指加在三相电压互感器一次绕组上的线电压,是绕组能够长期工作的电压,有 6、10、35、60、110、220、330、500 kV 等;接于三相系统与地之间的单相电压互感器,其一次额定电压为上述一次额定电压的 1/3。

选择时,电压互感器一次绕组额定电压应大于接入的被测电压的 0.9 倍,小于被测电压的 1.1 倍。即

$$0.9U_{1X} < U_{1e} < 1.1U_{1X}$$

(2)准确度等级的选择。电压互感器的准确度等级选择与电流互感器的准确度等级选择相同。

(3)接线方式的选择。电压互感器的接线方式有多种,计量有功电能和无功电能时,常用图 3-10 和图 3-11 两种接线方式。

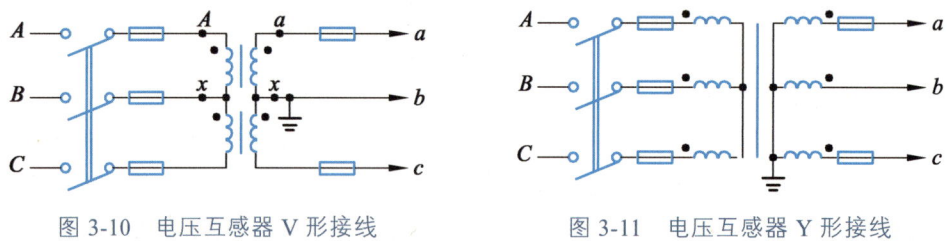

图 3-10　电压互感器 V 形接线　　　　图 3-11　电压互感器 Y 形接线

(4)额定容量的选择。按照二次负载取用的总视在功率 S 选择电压互感器的额定容量 S_e,公式为

$$0.25S_e < S < S_e$$

电压互感器每相的二次负载并不一定相等,因此应按最大一相取用的负载功率来考虑选择。

二次负载取用的总视在功率可按下式粗略计算

$$S = \sqrt{(\sum P)^2 + (\sum Q)^2}$$

式中　P——各仪表消耗的有功功率；

　　　Q——各仪表消耗的无功功率。

制造厂铭牌标定的额定二次负载通常用额定容量表示，其输出标准值有 10、15、25、30、50、75、100、150、200、250、300、400、500、1000 V·A。

对于三相电压互感器，由于互感器和负载接线方式不同，其二次负载容量的计算方法就不同。

2. 使用电压互感器应注意的问题

为了达到安全和准确测量的目的，使用电压互感器必须注意以下事项：

（1）按要求的相序进行接线，防止接错极性，否则将引起某一相电压升高 3 倍。

（2）电压互感器二次侧应可靠接地，以保证人身及仪表的安全。

（3）电压互感器二次侧严禁短路。

四、常用电压互感器的类型

电压互感器按相数分为单相、三相三芯柱式和三相五芯柱式；按绕组数分双绕组和三绕组。现简要介绍几种常用的电压互感器。

1. JDJ 型单相油浸双绕组电压互感器

这种电压互感器中，JDJ-6、JDJ-10 型为户内式，JDJ-35 型为户外式。图 3-12 为 JDJ-10 型单相电压互感器的外形及内部结构图。其结构简单，常用来测量线电压。

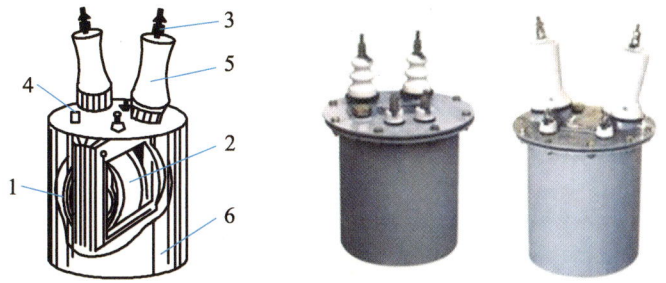

1—铁芯；2—绕组；3—二次绕组出线端；4—一次绕组出线端；5—套管绝缘子；6—外壳。

图 3-12　JDJ-10 型单相电压互感器

2. JSJW 型三相三绕组五柱式油浸电压互感器

它增加两个边柱铁心，构成五柱式，边柱可作为零序磁通的通路，如图 3-13 所示。如当系统 A 相发生单相接地短路时，电压互感器 A 相绕组被短接，B、C 相对地电压将升高 3 倍。根据对称分量法可知，在三相中将有零序电压和零序电流产生，零序电流在铁心中所产生零序磁通 ϕAO、ϕBO、ϕCO 通过两边柱铁心形成闭路，由于磁路磁阻很小，故零序电流值也小，发热少，不会危害电压互感器的安全运行。

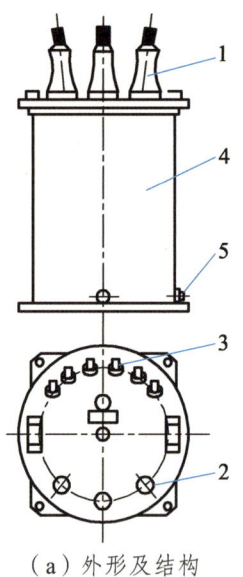
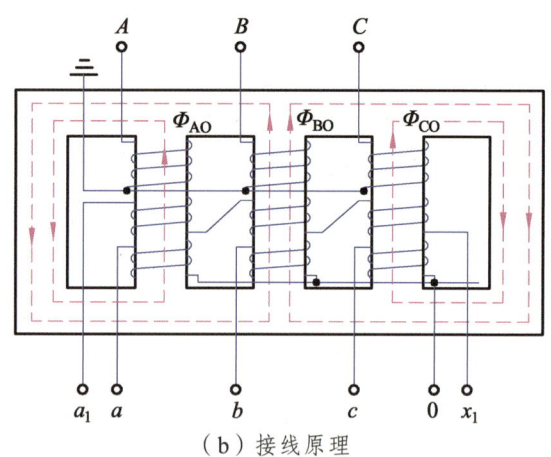

（a）外形及结构　　　　　　　（b）接线原理

（c）实物图

1—套管绝缘子；2—一次绕组出线端；3—二次绕组出线端；4—外壳；5—放油阀。

图 3-13　JSJW–10 型三相五柱式电压互感器

该型电压互感器有两个二次绕组：一个接成星形，供测量和继电保护用；另一个二次绕组也称辅助绕组，接成开口三角形，用来监视线路的绝缘情况。正常时开口三角形两端电压为 0 V，当一相接地时，开口三角形两端电压为 100 V。

3. JDZ 型电压互感器

JDZ 型电压互感器为单相双绕组环氧树脂浇注绝缘的户内用电压互感器。其优点：体积小，重量轻，节约铜和钢，能防潮、防盐雾、防霉，可用来代替 JDJ 型。

4. JDZJ 型电压互感器

JDZJ 型电压互感器为单相三绕组环氧树脂浇注绝缘的户内用电压互感器，可供中性点不直接接地系统测量电压、电能及单相接地保护用。其构造与 JDZ 型相似，其不同之处是 JDZJ 型有辅助二次绕组。使用时一次绕组的一端接高压，另一端接地，但其一次绕组两端均为全绝缘结构。一般 3 台 JDZJ 型电压互感器可代替一台 JSJW 型电压互感器。

Module 2　Voltage Transformer

3.2.1　Main Technical Data of Voltage Transformer

3.2.1.1　Classifications of voltage transformers

Currently, the classifications of voltage transformers are as follows according to different situations.

1. By purpose

Voltage transformer for measurement and voltage transformer for protection. These two types of voltage transformers can be divided into single-phase voltage transformer and three-phase voltage transformer.

2. By installation site

Indoor voltage transformer and outdoor voltage transformer.

3. By principle of voltage transformation

（1）Capacitive voltage transformer, using capacitive partial voltage to transform voltage.

（2）Photoelectric voltage transformer, using photoelectric elements to transform voltage.

（3）Electromagnetic voltage transformer, using electromagnetic induction to transform voltage.

Electromagnetic voltage transformer is the focus of this chapter. All voltage transformers without special instructions in the following text refer to electromagnetic voltage transformers.

4. By structure

（1）For single-pole voltage transformer, the primary and secondary windings are wound on the same iron core column.

（2）For cascade voltage transformer, the primary winding is divided into several sections with the same number of turns, and all sections are connected in series. One end is connected to the high-voltage circuit, and the other end is grounded.

3.2.1.2　Specification for models of voltage transformers

At present, the model arrangement method of domestic voltage transformer is shown as in Fig. 3-1.

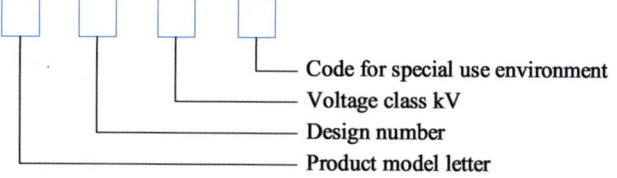

Fig. 3-1　The Model Arrangement Method of Domestic Voltage Transformer

The letters in the voltage transformer model are all represented by the Chinese phonetic alphabets, and the order of the letters and their corresponding symbolic meanings are as shown in Table 3-1.

The codes of voltage transformers in special use environment mainly include the following: CY—for use in ships; GY—for use in plateau areas; W—for use in dirty areas; AT—for use in dry and tropical areas; TH—for use in humid and tropical areas.

Table 3-1　Meaning and Arrangement Order of Voltage Transformer Model Letters

S/N	Category	Meaning	Letter
1	Name	Voltage transformer	J
2	Number of phases	Single phase	D
		Three-phase	S
3	Insulating medium outside the winding	Transformer oil	
		Air (dry)	G
		Cast into a solid shape	Z
		Gas	Q
4	Structural features	With backup voltage winding	X
		Three-column-core, with compensating winding	B
		Five-column-core, three windings per phase	W
		Cascade type, with backup voltage winding	C

Main parameters of voltage transformer are as follows:

1. Rated voltage of winding

The rated primary voltage refers to the voltage that can be applied to the primary winding for a long time, and its performance is determined under this reference. According to its connection to the circuit, it can be either line voltage or phase voltage. Its value shall be consistent with the "rated voltage" series stipulated by the power system of China.

According to relevant national regulations, the rated secondary voltage of single-phase voltage transformer connected between phases in the three-phase system is 100 V, and the rated secondary voltage of single-phase voltage transformer connected between the phase and the ground of the three-phase system is $100\sqrt{3}$ V.

2. Rated voltage transformation ratio

The rated voltage transformation ratio is the ratio of the rated primary voltage to the rated secondary voltage, which is generally expressed in the form of undivided fractions:

$$K_U = \frac{U_{1e}}{U_{2e}}$$

3. Rated secondary load

The rated secondary load of the voltage transformer is the secondary load admittance (or impedance) value based on which the accuracy class is determined. Rated output capacity refers to the capacity output at rated voltage when a rated load with a specified power factor is connected to the secondary circuit, usually expressed in terms of apparent power (V · A).

In the actual test, the secondary load of the voltage transformer is often expressed by the measured admittance, and the relationship between the load admittance and the output capacity is:

$$S = U_2^2 Y$$

Since the rated value of U_2 is 100 V, it can often be calculated by $S = Y \times 10^4$.

4. Accuracy class

Because there are certain errors in the voltage transformer, the accuracy class of the voltage transformer is divided according to the allowable errors of the voltage transformer. The accuracy classes of domestic voltage transformers are Class 0.2, Class 0.5, Class 1, and Class 3.

Voltage transformers of Class 0.2 are mainly used for precise measurement and metering in the laboratory, or as a standard to inspect low-level transformers, or can cooperate with standard instruments to inspect instruments; Voltage transformers of Class 0.5 are usually used for the measurement of power generation and distribution equipment; Voltage transformers of Class 1 and Class 3 are mainly used for non-precision measurement, monitoring, and protection. When the manufacturer marks the accuracy class on the nameplate, it must also indicate the secondary output capacity that determines the accuracy class.

5. Polarity mark

In order to ensure the correct wiring of measurement and calibration, the terminals of primary and secondary windings of voltage transformer shall be marked with polarity marks. The primary winding terminals of the voltage transformer are represented by uppercase letters A, B, C, and N, and the secondary winding terminals are represented by lowercase letters a, b, c, and n.

3.2.2 Working Principle and Error Characteristics

3.2.2.1 Working principle

The working principle, structure, and wiring method of the voltage transformer are similar to those of the power transformer. The voltage transformer is also composed of insulated primary and secondary windings wound on a common closed core, as shown in Fig. 3-2. The main difference is that the capacity of the two is different, and the voltage transformer works in a state close to no load.

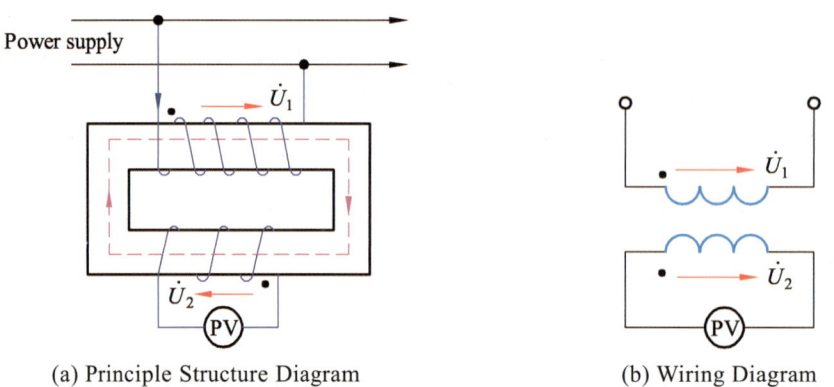

(a) Principle Structure Diagram   (b) Wiring Diagram

Fig. 3-2   Principle Structure Diagram and Wiring Diagram of Voltage Transformer

A voltage transformer changes the high voltage into low voltage to supply the instrument, so it has more primary turns N_1 and less secondary turns N_2. The primary winding is connected in parallel with the measured voltage, and the secondary winding is connected in parallel with the voltage coils of various measuring instruments or relays. The secondary side of voltage transformer shall be equipped with fuse to protect itself from damage due to short circuit of secondary winding; Where possible, fuses shall also be installed on the primary side to protect the high voltage power network from endangering the safety of the primary system due to the transformer primary winding or lead failure. The voltage transformer is represented by TV/PT in the electrical diagram.

When the voltage is applied to the primary winding, there is an alternating main flux in the core, and the primary and secondary windings have induced electromotive force $\dot{E}_1$ and $\dot{E}_2$, respectively. After converting the impedance of the secondary winding of the voltage transformer to the primary side, the T-shaped equivalent circuit diagram and phasor diagram as shown in Fig. 3-3 and Fig. 3-4 can be obtained.

From the equivalent circuit diagram

$$\dot{U}_1 = \dot{I}_1(R_1 + jX_1) - \dot{E}_1$$
$$\dot{U}_2' = \dot{E}_2' - \dot{I}_2'(R_2' + jX_2')$$

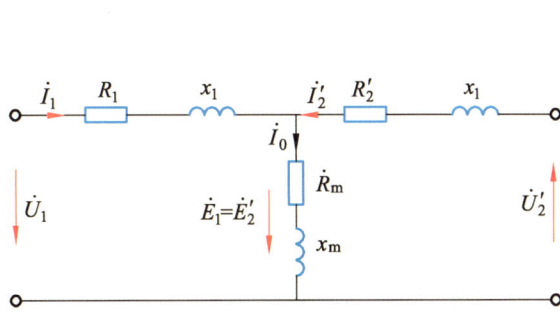
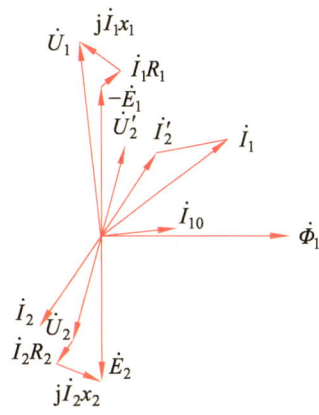

Fig. 3-3 T-shaped Equivalent Circuit Diagram of Voltage Transformer

Fig. 3-4 Phasor Diagram of Voltage Transformer

Where, R_1, X_1—resistance and impedance of the primary winding;

R'_2, X'_2—resistance and impedance of the secondary winding converted to the primary side.

If the voltage drop generated by the excitation current and load current in the primary and secondary windings is ignored, then

$$K_U = \frac{U_1}{U_2} = \frac{E_1}{E_2} = \frac{N_1}{N_2}$$

This is the voltage transformation ratio of the ideal voltage transformer, called the rated transformation ratio, that is, the ratio of the ideal voltage transformer primary winding voltage U_1 to the secondary winding voltage U_2 is a constant, equal to the ratio of the number of turns of the primary winding and the secondary winding.

In fact, voltage transformer is subject to iron loss and copper loss, and there is impedance voltage drop in the winding. It can be seen from phasor diagram 3-4 that after the secondary voltage rotates 180° ($-\dot{U}'_2$), it is different from the primary voltage $\dot{U}_1$, and there is phase difference, which means that there are ratio difference and angle difference in voltage transformer.

The ratio difference is expressed in f_U, which is equal to

$$f_U = \frac{U'_2 - U_1}{U_1} \times 100\% = \frac{\frac{N_1}{N_2}U_1 - U_1}{U_1} \times 100\% = \frac{K_U - K'_U}{K'_U} \times 100\%$$

Where, U_1—actual primary voltage effective value;

U_2—actual secondary voltage effective value;

K'_U—actual voltage transformer transformation ratio, $K'_U = \frac{U_1}{U_2}$;

K_U—rated voltage transformer transformation ratio, $K_U = \frac{U_{1e}}{U_{2e}} = \frac{N_1}{N_2}$.

The phase angle difference, abbreviated as angle difference, refers to the phase difference between the primary voltage and the secondary voltage phasor after 180 degrees of rotation, expressed as δ_U, the unit is " ' ". When the secondary voltage after rotation leads the phasor of the primary voltage, the angle difference is positive; Otherwise, the angle difference is negative.

3.2.2.2 Error characteristics

1. Load characteristics of voltage transformer

The magnitude and nature of the secondary load current of the voltage transformer affect the ratio difference and angle difference. Fig. 3-5 shows the relationship between the ratio difference and angle difference as a function of the secondary current when the primary voltage $\dot{U}_1$ remains constant. For no load ($I_2=0$) in the figure, the ratio difference is negative and the angle difference is positive. The ratio difference and angle difference during no-load operation are determined by the structure of the transformer and the no-load current. The larger the no-load current, the more the ratio difference increases in the negative direction and the angle difference increases in the positive direction.

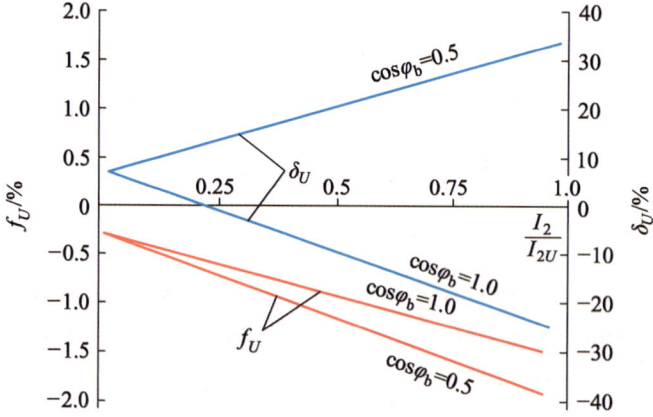

Fig. 3-5 Load Characteristics of Voltage Transformer

With the increase of load current, the ratio difference continues to increase in the negative direction on the basis of no-load, and the lower the power factor is, the more it increases in the negative direction. When the power factor is low, the angle difference always increases in a positive direction with the increase of load current. When the power factor is high, it first changes from a positive value to zero and then increases in the negative direction.

The effect of secondary load on the ratio difference and angle difference is caused by the voltage drop generated by the secondary current in the winding, therefore, limiting the current density of the winding wires and reducing the magnetic flux leakage of the windings to reduce the leakage reactance are effective measures to improve the accuracy of the voltage transformer.

2. Voltage characteristics of voltage transformer

The voltage characteristic is the relationship between the ratio difference and angle difference of the voltage transformer and the primary voltage, and its changing trend is as shown in Fig. 3-6. The effect of primary voltage on the ratio difference and angle difference can be explained by the magnetization curve of the core. As shown in Fig. 3-7, because the magnetic conductivity μ and the loss angle α of the core are nonlinear, for the same magnetic field intensity increment ΔH, the magnetic induction intensity increment ΔB corresponding to different parts of the magnetization curve is different. When the primary voltage is low, because the magnetic conductivity μ is low, the required excitation current I_0 is large, so the ratio difference f_0 and angle difference δ_0 are large in no-load. With the increase of voltage, the magnetic conductivity increases, and the core works in the flat part of the magnetization curve, so the ratio difference and angle difference begin to decrease and tend to stabilize gradually. It can be seen that the primary side of the voltage transformer shall operate under the rated voltage.

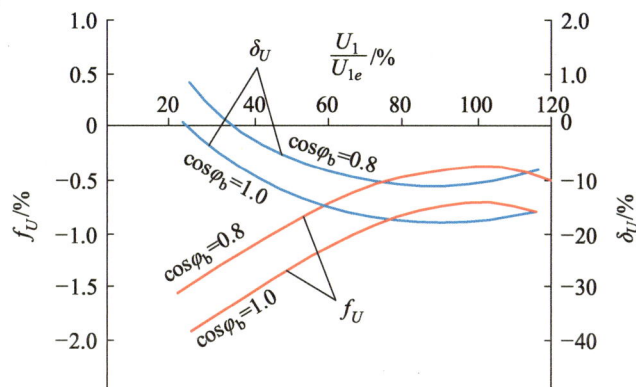

Fig. 3-6　Voltage Characteristics of Voltage Transformer

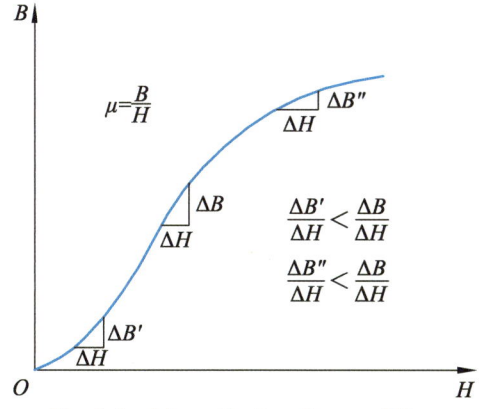

Fig. 3-7　Magnetization Curve of Core

3.2.2.3　Compensation method of voltage transformer error

As mentioned above, the voltage drop caused by the no-load current and the load current on the impedance of the primary and secondary windings of the voltage transformer, especially on the resistance, is the main cause of the ratio difference and angle difference. The greater the I_0, the greater the no-load error (also known as the inherent error). The larger the internal parameters (R_1, R_2, X_1, X_2), the greater the load current will cause a significant voltage drop and increase the error. In addition, the change of the ratio of R_1 and R_2 to X_1 and X_2 also has an effect on the angle difference. Therefore, in order to improve the accuracy of voltage transformer, the following measures can be taken in terms of design.

(1) The core shall be materials with high magnetic conductivity, such as single-crystal cold-rolled silicon steel sheet, permalloy alloy containing 50% nickel, and so on.

(2) Properly increase the cross-sectional area of the winding wire to reduce the resistance.

(3) The winding assembly shall be as compact and uniform as possible to reduce leakage reactance.

The above measures inevitably increase manufacturing costs. For this reason, the following error compensation methods can also be adopted.

1. Turn compensation method

Because the primary and secondary voltage of the voltage transformer is proportional to the number of turns of the winding, if the number of turns of the winding is increased or decreased appropriately, the voltage will increase or decrease accordingly. For example, if the secondary winding is appropriately increased, the secondary voltage increases. When the primary voltage remains unchanged, the actual transformation ratio K_U decreases. If the number of turns in the secondary winding is reduced properly, the ratio difference will change to a negative direction and the characteristic of the ratio difference will become worse. The effect of changing the number of turns in the primary winding on the ratio difference is opposite to that of changing the number of turns in the secondary winding.

In order not to adjust too much, the secondary winding can be wound with two or more strands of wire. When double-stranded or n-stranded conductors are used to increase (or decrease) the number of turns, the compensation effect of 1/2 or $1/n$ turns can be achieved. In the actual adjustment, in order to obtain the required ratio difference characteristics, the ratio difference is generally adjusted to zero at 50% rated load. In this way, the ratio difference is positive when the load is small and negative when the load is large. Therefore, the ratio difference characteristics can be improved.

It shall be pointed out that the turn compensation method can only improve the characteristics of ratio difference, but not the characteristics of angle difference.

2. Additional winding compensation method

As shown in Fig. 3-8 (a), an additional winding (also known as compensation winding) N_K is connected in parallel to the secondary winding N_2 by resistor R. By making N_K one or more turns less than N_2, so they produce induced electromotive forces $\dot{E}_K < \dot{E}_2$, but the phase of the two electromotive forces is the same. The electromotive force difference between them is $\Delta \dot{E}$, a balanced current $\dot{I}_P$ is generated under the action of $\Delta \dot{E}$, as shown in Fig. 3-8 (b), the amplitude and phase of the $\dot{I}_P$ are determined by the resistance R and the internal impedance of the two windings N_K and N_2. Therefore, by changing the number of turns of resistors R, N_K, and N_2, the magnitude and phase of $\dot{I}_P$ can be changed, resulting in a change in the magnitude and phase of the secondary voltage $\dot{U}_2$, and a change in the ratio difference f_U and angle difference δ_U accordingly. As long as the number of turns of N_K and N_2 and the value of R are properly selected, f_U and δ_U can be compensated to a satisfactory degree.

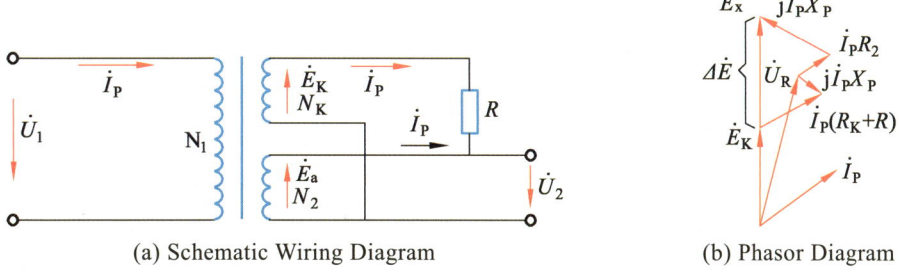

(a) Schematic Wiring Diagram (b) Phasor Diagram

Fig. 3-8 Compensation Schematic Diagram of Parallel Additional Winding

3. Compensation method of external impedance of secondary winding in parallel

Usually, parallel capacitors are used for compensation, as shown in Fig. 3-9. After parallel capacitor C is connected, a capacitor current $\dot{I}_C$ that is 90 degrees ahead of its secondary voltage is generated in the secondary winding. According to the principle of magnetomotive force balance, a current $-\dot{I}_C$ must be added to the primary winding in order to balance with $\dot{I}_C$ phase. $\dot{I}_C$ When and $-\dot{I}_C$ flow through the primary and secondary windings respectively, the impedance voltage drop $\dot{I}_C Z_2$ and $-\dot{I}_C Z_2$ will be produced, so it will cause additional errors. If $\dot{I}_C Z_2$ is converted to $\dot{I}'_C Z'_2$ on the primary side, and the additional error caused by the capacitive current is expressed as a complex error, there is

$$\Delta \dot{\gamma}_U = -\frac{\dot{I}'_C (Z_1 + Z'_2)}{\dot{U}_1}$$

As shown in Fig. 3-9 (b), the horizontal component of $\Delta \dot{\gamma}_U$ is the additional ratio difference $\Delta \dot{f}_U$, which can compensate for $\dot{f}_U$; $\Delta \dot{\gamma}_U$ The vertical component of is additional angle difference $\Delta \dot{\delta}_U$, which has no compensation effect on $\dot{\delta}_U$. The compensation value is proportional to the internal impedance and capacitive current of the primary and secondary windings, but independent of the secondary load, so the parallel capacitor compensation method actually only compensates the no-load error.

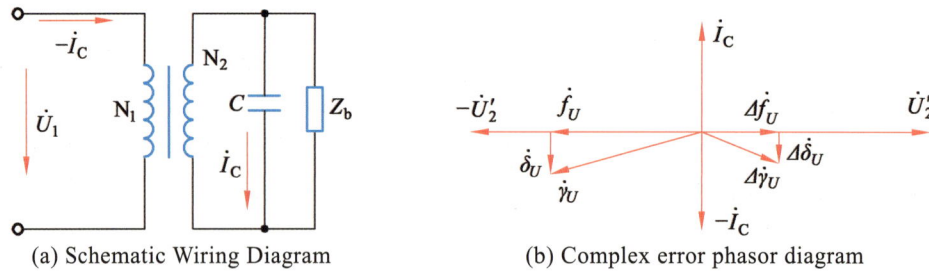

(a) Schematic Wiring Diagram   (b) Complex error phasor diagram

Fig. 3-9   Compensation Principle of Parallel External Impedance

3.2.3 Correct Use of Voltage Transformer

1. Selection of voltage transformer

(1) Selection of rated voltage. The rated voltage of the voltage transformer refers to the line voltage applied to the primary winding of the three-phase voltage transformer, which is the voltage under which the winding can work for a long time, such as 6 kV, 10 kV, 35 kV, 60 kV, 110 kV, 220 kV, 330 kV, and 500 kV; The primary rated voltage of the single-phase voltage transformer connected between the three-phase system and the ground is $1/\sqrt{3}$ of the above-mentioned rated voltage.

When selecting, the rated voltage of the primary winding of the voltage transformer shall be greater than 0.9 times the connected measured voltage and less than 1.1 times the measured voltage. i.e.

$$0.9U_{1X} < U_{1e} < 1.1U_{1X}$$

(2) Selection of accuracy class. The accuracy class selection of voltage transformer is the same as that of current transformer.

(3) Selection of wiring method. There are many wiring methods of voltage transformer. When measuring active energy and reactive energy, two common wiring methods are shown in Fig. 3-10 and Fig. 3-11.

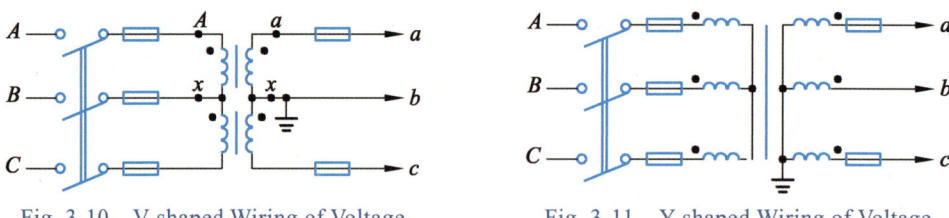

Fig. 3-10   V-shaped Wiring of Voltage Transformer

Fig. 3-11   Y-shaped Wiring of Voltage Transformer

(4) Selection of rated capacity. Select the rated capacity S_e of the voltage transformer according to the total apparent power S of the secondary load. The formula is as follows:

$$0.25S_e < S < S_e$$

The secondary load of each phase of the voltage transformer is not necessarily equal, so it shall be selected according to the load power of the maximum one phase.

The total apparent power of secondary load can be roughly calculated according to the following formula:

$$S = \sqrt{(\sum P)^2 + (\sum Q)^2}$$

Where, P—active power consumed by each instrument;

Q—reactive power consumed by each instrument.

The rated secondary load marked on the manufacturer nameplate is usually expressed by the rated capacity, and its output standard values are 10, 15, 25, 30, 50, 75, 100, 150, 200, 250, 300, 400, 500, and 1000 V · A.

For the three-phase voltage transformer, the calculation method of the secondary load capacity is different because of the different wiring methods of the transformer and the load.

2. Matters needing attention in the use of voltage transformer

In order to measure safely and accurately, attention must be paid to the following items when using voltage transformer:

(1) Conduct wiring according to the required phase sequence to prevent incorrect polarity, otherwise, it will cause a voltage increase of   times in a certain phase.

(2) The secondary side of the voltage transformer shall be reliably grounded to ensure the safety of the person and instrument.

(3) Short circuit at the secondary side of voltage transformer is strictly prohibited.

3.2.4　Types of Commonly Used Voltage Transformers

By number of phases, voltage transformers are divided into single-phase, three-phase three-core-column and three-phase five-core-column ones; By number of windings, voltage transformers are divided into double-winding and three-winding ones. Several commonly used voltage transformers are briefly introduced below.

1. JDJ single-phase oil-immersed double-winding voltage transformer

Among this type of voltage transformer, JDJ-6 and JDJ-10 are indoor, and JDJ-35 is outdoor. Fig. 3-12 shows the outline and internal structure of JDJ-10 single-phase voltage transformer. It is of simple structure and is often used to measure line voltage.

(1) Outline and structure　　　　　　(2) Picture

1—core; 2—Winding ; 3—Secondary winding outlet terminal; 4—Primary winding outlet terminal; 5—Bushing insulator; 6—External.

Fig. 3-12　JDJ-10 Single-phase Voltage Transformer

2. JSJW three-phase three-winding five-column oil-immersed voltage transformer

It has two additional side column cores to form a five-column type, and the side column can be used as the path of zero sequence flux, as shown in Fig. 3-13. For example, when a single-phase grounding short circuit occurs in phase A of the system, the phase A winding of the voltage transformer is short-connected, and the relative ground voltage of B and C will be increased by $\sqrt{3}$ times. According to the symmetrical component method, zero sequence voltage and zero sequence current will be generated in the three phases, and the zero sequence magnetic fluxes ϕAO, ϕBO, and ϕCO produced by the zero sequence current in the core form a closed circuit through the cores on both sides of the column. Because the magnetoresistance of the magnetic circuit is very small, so the zero sequence current value is also small, the heating is less, and it will not endanger the safe operation of the voltage transformer.

This type of voltage transformer has two secondary windings, one of which is connected into a star for measurement and relay protection. Another secondary winding, also known as an auxiliary winding, is connected into an open triangle to monitor the insulation condition of the circuit. Normally, the voltage at both ends of the open triangle is 0 V, and when one phase is grounded, the voltage at both ends of the open triangle is 100 V.

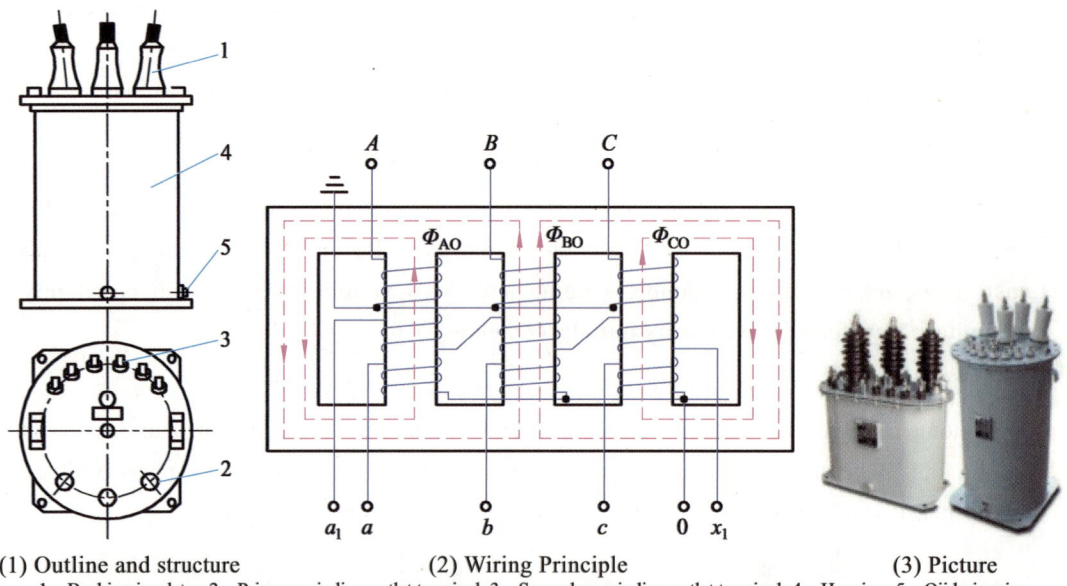

(1) Outline and structure　　　　(2) Wiring Principle　　　　(3) Picture
1—Bushing insulator; 2—Primary winding outlet terminal; 3—Secondary winding outlet terminal; 4—Housing; 5—Oiidrainvaive.

Fig. 3-13　JSJW-10 Three-phase Five-column Voltage Transformer

3. JDZ voltage transformer

JDZ voltage transformer is an indoor voltage transformer with single-phase double-winding epoxy resin casting insulation. Its advantages: small size, light weight, saving copper and steel, moisture-proof, salt fog-proof, mildew-proof, can be used to replace the JDJ model.

4. JDZJ voltage transformer

JDZJ voltage transformer is an indoor voltage transformer with single-phase three-winding epoxy resin casting insulation, which can be used for measuring voltage, electric energy, and single-phase grounding protection of the system where neutral point is not directly grounded. Its structure is similar to that of JDZ model, except that JDZJ model has auxiliary secondary winding. When in use, one end of the primary winding is connected to high voltage and the other end is grounded, but both ends of the primary winding are fully insulated. Generally, three JDZJ voltage transformers can replace one JSJW voltage transformer.

模块三　电流互感器

一、电流互感器的主要技术数据

（一）电流互感器分类

目前，电流互感器的分类按不同情况划分如下。

（1）电流互感器按用途可分为两类：一类是测量电流、功率和电能用的测量用互感器；另一类是继电保护和自动控制用的保护控制用互感器。

（2）根据一次绕组匝数可分为单匝式和多匝式，如图 3-14 所示。单匝式又分为贯穿型和母线型两种。贯穿型互感器本身装有单根铜管或铜杆作为一次绕组；母线型互感器则本身未装一次绕组，而是在铁心中留出一次绕组穿越的空隙，施工时以母线穿过空隙作为一次绕组。通常油断路器和变压器套管上的装入式电流互感器就是一种专用母线型互感器。

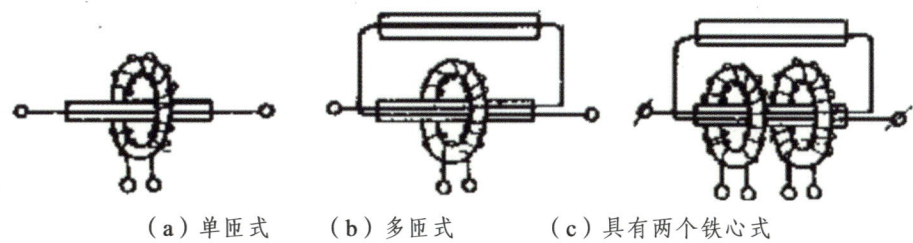

（a）单匝式　　（b）多匝式　　（c）具有两个铁心式

图 3-14　电流互感器的结构原理

（3）根据安装地点可分为户内式和户外式。

（4）根据绝缘方式可分为干式，浇注式，油浸式等。干式用绝缘胶浸渍，适用于作为低压户内的电流互感器；浇注式用环氧树脂作绝缘，浇注成型；油浸式多为户外型。

（5）根据电流互感器工作原理可分为电磁式、光电式、磁光式、无线电式电流互感器。

（二）电流互感器的型号规定

目前，国产电流互感器型号编排方法规定如下：

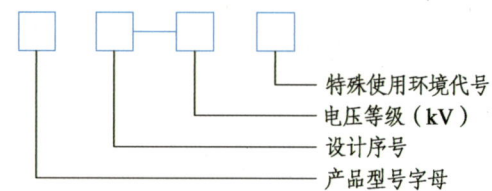

图 3-15　国产电流互感器型号示意图

产品型号均以汉语拼音字母表示，字母含义及排列顺序见表 3-2。

表 3-2　电流互感器型号字母含义

第一个字母		第二个字母		第三个字母		第四个字母		第五个字母	
字母	含义	字母	含义	字母	含义	字母	含义	字母	含义
L	电流互感器	A	穿墙式	C	瓷绝缘	B	保护级	D	差动保护
		B	支持式	G	改进的	D	差动保护		
		C	瓷箱式	J	树脂浇注	J	加大容量		
		D	单匝式	K	塑料外壳	Q	加"强"式		
		F	多匝式	L	电容式绝缘	Z	浇注绝缘		
		J	接地保护	M	母线式				
		M	母线式	P	中频				
		Q	线圈式	S	速饱和				
		R	装入式	W	户外式				
		Y	低压的	Z	浇注绝缘				
		Z	支柱式						

（三）电流互感器的主要参数

1. 额定电流变比

额定电流变比是指一次额定电流与二次额定电流之比（有时简称电流比）。额定电流比一般用不约分的分数形式表示，如一次额定电流 I_{1e} 和二次额定电流 I_{2e} 分别为 100 A、5 A，则

$$K_I = I_{1e} / I_{2e} = 100/5$$

所谓额定电流，就是在这个电流下，互感器可以长期运行而不会因发热损坏。当负载电流超过额定电流时，叫作过负载。如果互感器长期过负载运行，会把它的绕组烧坏或缩短绝缘材料的寿命。

2. 准确度等级

由于电流互感器存在着一定的误差，因此根据电流互感器允许误差划分互感器的准确度等级。国产电流互感器的准确度等级有 0.2s 级、0.5s 级、0.2 级、0.5 级、5 级、10 级。0.2s 级电流互感器，主要用于试验室进行精密测量，或者作为标准用来检验低等级的互感器，也可以与标准仪表配合，用来检验仪表，所以也叫作标准电流互感器。用户电能计量装置通常采用 0.2 级和 0.5 级电流互感器，保护与控制可采用 5 级和 10 级的电流互感器。

3. 额定容量

电流互感器的额定容量，就是额定二次电流 I_{2e} 通过二次额定负载 Z_{2e} 时所消耗的视在功率 S_{2e}，所以

$$S_{2e} = I_{2e}^2 \times Z_{2e}$$

一般情况 $I_{2e} = 5 A$，因此，额定容量也可以用额定负载阻抗 Z_{2e} 表示。

电流互感器在使用中，二次连接线及仪表电流线圈的总阻抗，不能超过铭牌上规定的额定容量且不低于 1/4 额定容量时，才能保证它的准确度。制造厂铭牌标定的额定二次负载通常用额定容量表示，其输出标准值有 5 V·A、10 V·A、15 V·A、25 V·A、30 V·A、50 V·A、100 V·A 等。

4. 额定电压

电流互感器的额定电压，是指一次绕组长期对地能够承受的最大电压（有效值）。它只是说明电流互感器的绝缘强度，而和电流互感器额定容量没有任何关系。它标在电流互感器型号后面。例如 LCW-35，其中"35"是指额定电压，它以 kV 为单位。

5. 极性标志

为了保证测量及校验工作的接线正确，电流互感器一次和二次绕组的端子应标明极性标志。

（1）一次绕组首端标为 L_1，末端标为 L_2。当多量限一次绕组带有抽头时，首端标为 L_1，自第一个抽头起依次标为 L_2，L_3……。

（2）二次绕组首端标为 k_1，末端标为 k_2。当二次绕组带有中间抽头时，首端标为 k_1，自第一个抽头起以下依次标志为 k_2，k_3……。

（3）对于具有多个二次绕组的电流互感器，应分别在各个二次绕组的出线端标志"k"前加注数字，如 $1k_1$，$1k_2$，$1k_3$……；$2k_1$，$2k_2$，$2k_3$……。

（4）标志符号的排列应当使一次电流自 L_1 端流向 L_2 端时，二次电流自 k_1 流出，经外部回路流回到 k_2。

从电流互感器一次绕组和二次绕组的同极性端子来看,电流 I_1、I_2 的方向是相反的,这样的极性关系称为减极性,反之称为加极性。电流互感器一般都按减极性表示。

二、电流互感器的结构和工作原理

(一)电流互感器的结构

目前,电力系统中使用的电流互感器一般为电磁式,其基本结构与一般变压器相似,由两个绕制在闭合铁心上、彼此绝缘的绕组(一次绕组和二次绕组)所组成,其匝数分别为 N_1 和 N_2,如图 3-16 所示。一次绕组与被测电路串联,二次绕组与各种测量仪表或继电器的电流线圈相串联。

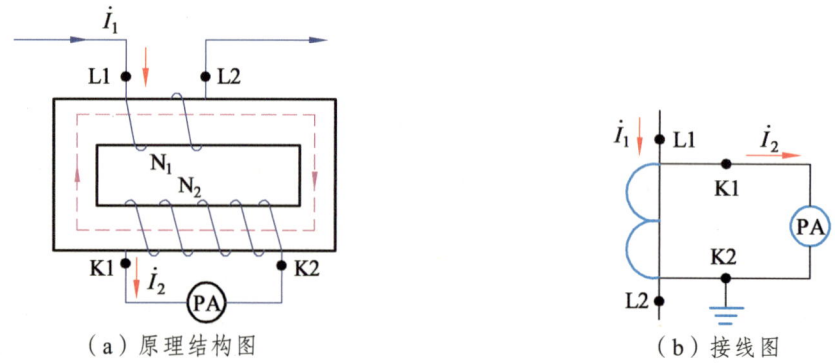

图 3-16 电流互感器原理结构图和接线图

电力系统中,经常将大电流 I_1 变为小电流 I_2 进行测量,所以二次绕组的匝数 N_2 大于一次绕组的匝数 N_1。电流互感器的二次额定电流一般为 5A,也有 1A 和 0.5A 的。电流互感器在电气图中文字符号用 TA 表示。

(二)工作原理和特性

电流互感器的工作原理与一般变压器的工作原理基本相同。当一次绕组中有电流 $\dot{I}_1$ 通过时,一次绕组的磁动势 $\dot{I}_2N_1$ 产生的磁通绝大部分通过铁心而闭合,从而在二次绕组中感应出电动势 $\dot{E}_2$。如果二次绕组接有负载,那么二次绕组中就有电流 $\dot{I}_2$ 通过,有电流就有磁动势,所以二次绕组中由磁动势 I_2N_2 产生磁通,这个磁通绝大部分也是经过铁心而闭合。因此铁心中的磁通是由一、二次绕组的磁动势共同产生的合成磁通 $\dot{\Phi}$,称为主磁通。根据磁动势平衡原理可以得到

$$\dot{I}_1N_1 + \dot{I}_2N_2 = \dot{I}_{10}N_1$$

式中,$\dot{I}_{10}N_1$——励磁磁动势。

如果忽略铁心中各种损耗,可认为 $\dot{I}_{10}N \approx 0$,则

$$\dot{I}_1N_1 + \dot{I}_2N_2 = 0$$

$$\dot{I}_1N_1 = -\dot{I}_2N_2$$

这是理想电流互感器的一个很重要的关系式，即一次磁动势安匝等于二次磁动势安匝，且相位相反。进一步化简上式，得到

$$K_1 = \frac{I_{1e}}{I_{2e}} = \frac{N_2}{N_1}$$

即理想电流互感器两侧的额定电流大小和它们的绕组匝数成反比，并且等于常数 K_1，称为电流互感器的额定变比。

电流互感器的基本工作原理、结构型式与普通变压器相似，但是电流互感器的工作状态与普通变压器有显著的区别：

（1）电流互感器的一次电流（I_1）取决于一次电路的电压和阻抗，与电流互感器的二次负载无关，即当二次负载变化时，例如多串几只电流表或少串几只电流表，不能改变其一次电流值的大小。

（2）电流互感器二次电路所消耗的功率随二次电路阻抗的增加而增大，即

$$S_2 = I_2^2 \times Z_2$$

（3）电流互感器二次电路的负载阻抗都是些内阻很小的仪表，如电流表以及电能表的电流线圈等，所以其工作状态接近于短路状态。

普通电流互感器的铁心通常制成芯式，材料是优质硅钢片。为了减小涡流损耗，片与片之间彼此绝缘。准确度级别高的实验室用电流互感器铁心是用坡莫合金制成，其截面为环形，这种合金具有较高的起始导磁率以及很小的损耗。

三、电流互感器的接线方式

1. 两相星形（V形）连接

由两台电流互感器构成，A 和 C 相所接电流互感器的二次绕组一端接到表计，另一端相互连接后至 B 相表计或接至 a、c 相表计出线端连接处。两台电流互感器的二次绕组电流分别为 $\dot{I}_a$ 和 $\dot{I}_c$，公共接线中流过的电流为 $\dot{I}_b = -(\dot{I}_a + \dot{I}_c)$，如图 3-17 所示，这种连接方式常用在三相三线电路中。它的优点：

（1）节省导线。

（2）能利用接线方法取得第三相电流，一般为 B 相电流。

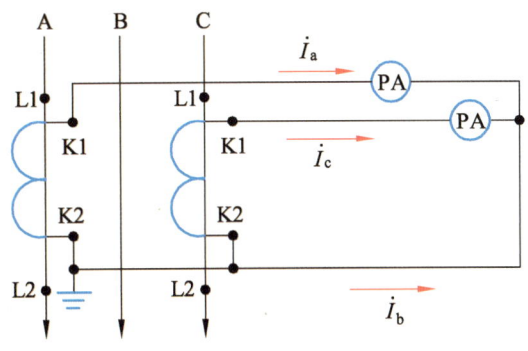

图 3-17 两相星形（V形）原理接线图

但这种连接方法有其缺点：

（1）现场用单相方法校验时，由于实际二次负载与运行时不一致，有时必须要采用三相方法（或其他类似方法），给校验工作带来一些困难。

（2）由于有可能其中一相极性接反，公共线电流变成差电流，使错误接线机率相对地较多一些。为此，有的地区在电能计量回路中采用分相接法。

2. 分相连接

分相连接就是各相分别连接，如图 3-18 所示。其优点是：
（1）现场校验与实际运行时负载相同。
（2）错误接线机率相对地少些。缺点是增加了一根导线。

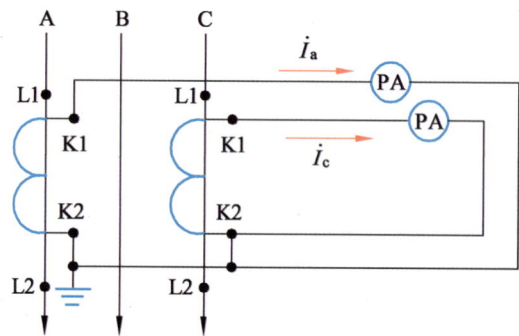

图 3-18　分相原理接线图

3. 三相星形（Y形）连接

三相四线电路中多采用三相星形连接，如图 3-19 所示。图中，A、B、C 三相电流互感器的二次绕组分别流过电流 $\dot{I}_a$、$\dot{I}_b$、$\dot{I}_c$。当三相电流不平衡时，公共接线中的电流 $\dot{I}_N = \dot{I}_a + \dot{I}_b + \dot{I}_c$，当三相电流平衡时，$\dot{I}_N = 0$。这种接线方法不允许断开公开接线，否则影响计量精度（因为零序电流没有通路）。

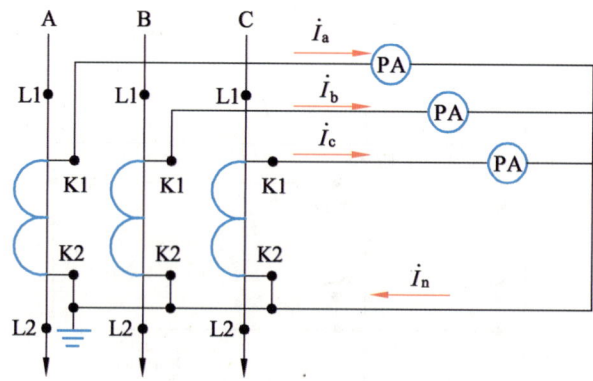

图 3-19　三相星形（Y形）原理接线图

四、常用的电流互感器

1. LFC-10 型多匝穿墙式电流互感器

如图 3-20、3-21 所示,其一次绕组穿过瓷绝缘套管,并固定在法兰盘上,两端附有接头盒,一次绕组引出端接线板与配电装置母线连接。二次绕组装在封闭的外壳内,二次绕组由接线端子引出。

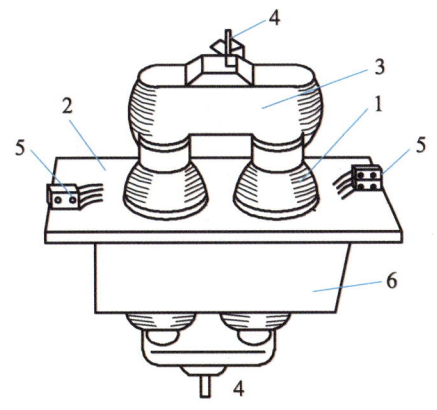
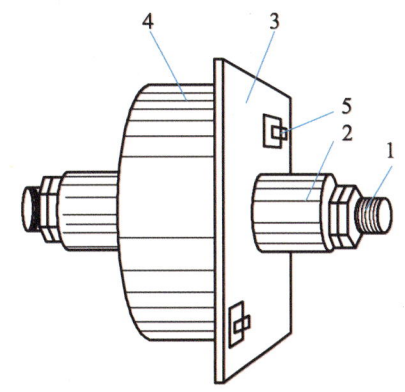

1—一次绕组;2—瓷套管;3—法兰盘;4—接线板;
5—二次绕组接线端子;6—外壳。

图 3-20　LFC-10 型电流互感器 1

1—瓷绝缘套管;2—法兰盘;3—接头盒;4—外壳;
5—二次绕组接线端子。

图 3-21　LFC-10 型电流互感器 2

2. LDC-10 单匝穿墙式电流互感器

这种系列的电流互感器如图 3-22 和图 3-23 所示,一次绕组为贯穿于瓷件中的一根紫铜棒,一次绕组和瓷件贯穿在铁心中,固定在法兰盘上,二次绕组装在封闭外壳内,二次绕组由引出端子引出。

图 3-22　LDC-10 电流互感器实物图 1

图 3-23　LDC-10 电流互感器实物图 2

3. LQJ-10 型环氧树脂浇注电流互感器

该型电流互感器如图 3-24 所示,一次和二次绕组由环氧树脂浇注,一次绕组由引出端引出，二次绕组由接线端引出。其主要优点是体积小，质量轻，电气绝缘性能好。目前用于 10 kV 及以下配电装置。LQJ-10 电流互感器实物见图 3-25。

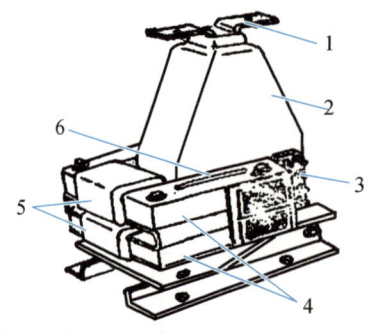

1—一次接线瑞；2—一次绕组环氧树脂浇注；3—二次接线端；4—铁心；
5—二次绕组；6—警告牌（上写"二次绕组不得开路"等字样）。

图 3-24　LQJ-10 电流互感器　　　　图 3-25　LQJ-10 电流互感器实物图

4. LMC-10 母线型穿墙式电流互感器

它的一次绕组为母线，穿过瓷套管，瓷套管又贯穿在铁心中，瓷套管两端有瓷套帽和夹板，为夹紧母线之用，如图 3-26 所示。LMC-10 电流互感器实物见图 3-27。

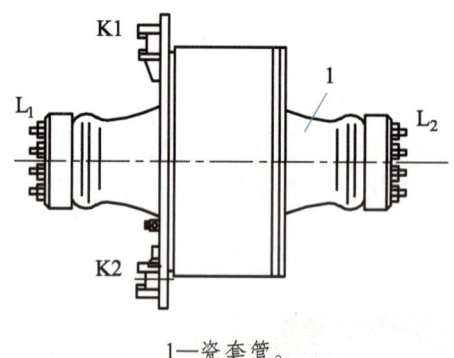
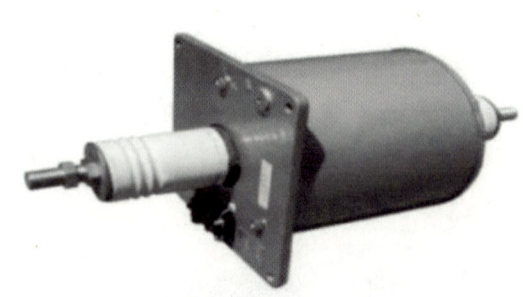

1—瓷套管。

图 3-26　LMC-10 电流互感器　　　　图 3-27　LMC-10 电流互感器实物图

Module 3　Current Transformer

3.3.1　Main Technical Data of Current Transformer

3.3.1.1　Classifications of current transformers

Currently, the classifications of current transformers are as follows according to different situations.

(1) Current transformers can be divided into two categories according to their purpose: one is for measuring current, power, and electrical energy; The second is the protection and control transformer used for relay protection and automatic control.

(2) According to the number of turns in the primary winding, current transformers can be divided into single-turn type and multi-turn type, as shown in Fig. 3-14. Single-turn type is divided into two types: through type and bus type. The through type transformer itself is equipped with a single copper tube or rod as a primary winding; The bus type transformer itself is not equipped with a primary winding, but rather leaves a gap for the primary winding to pass through in the core. During construction, the bus passes through the gap as the primary winding. Usually, the installed current transformer on the oil circuit breaker and transformer bushing is a special bus transformer.

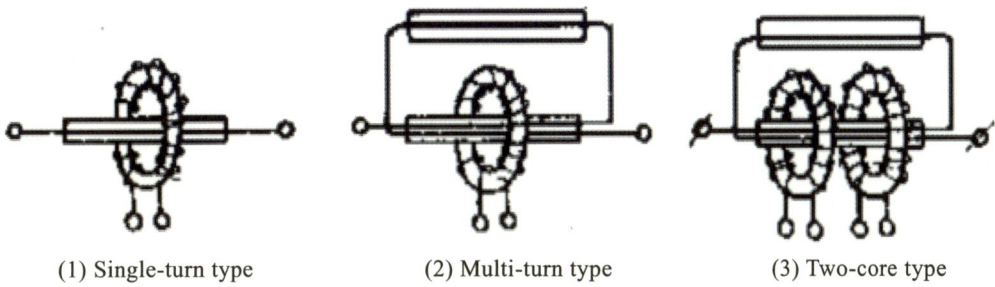

(1) Single-turn type　　　(2) Multi-turn type　　　(3) Two-core type

Fig. 3-14　Structure Principle of Current Transformer

(3) It can be divided into indoor type and outdoor type according to the installation site.

(4) According to the insulation mode, it can be divided into dry, pouring type, and oil immersed type. The dry type is impregnated with insulating glue, which is suitable for low-voltage indoor current transformer; The pouring type is insulated with epoxy resin and formed by pouring. The oil immersed type is mostly outdoor type.

(5) According to the working principle of current transformer, it can be divided into electromagnetic type, photoelectric type, magneto-optical type, and radio type.

3.3.1.2　Specification for models of current transformers

At present, the model arrangement method of domestic current transformer is as follows:

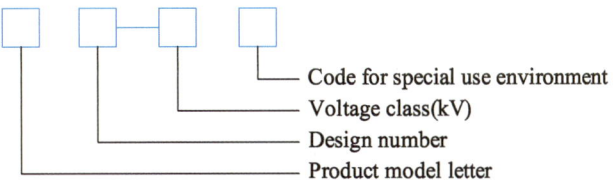

Fig. 3-15　Domestic Current Transformer

Models are all represented by the Chinese phonetic alphabets, and the meanings and orders of the letters are as shown in Table 3-2.

Table 3-2 Meaning of Current Transformer Model Letters

The First Letter		The Second Letter		The Third Letter		The Fourth Letter		The Fifth Letter	
Letter	Meaning	Letter	Meaning	Letter	Meaning	Letter	Meaning	Letter	Meaning
L	Current transformer	A	Through-the-wall	C	Porcelain insulator	B	Protection level	D	Differential protection
		B	Supported	G	Improved	D	Differential protection		
		C	Porcelain box type	J	Resin cast	J	Increased capacity		
		D	Single-turn type	K	Plastic enclosure	Q	"Reinforced" type		
		F	Multi-turn type	L	Capacitive insulation	Z	Cast insulation		
		J	Grounding protection	M	Bus type				
		M	Bus type	P	Intermediate frequency				
		Q	Coil type	S	Quick saturation				
		R	Built-in type	W	Outdoor type				
		Y	Low voltage	Z	Cast insulation				
		Z	Post type						

3.3.1.3 Major Parameters for Current Transformer

1. Rated current transformation ratio

Rated current transformation ratio refers to the ratio between primary rated current and secondary rated current (it is sometimes referred to as current ratio). The rated current is normally expressed in an irreducible fraction form. For instance, where primary rated current I_{1e} and secondary rated current I_{2e} are 100 A and 5 A respectively,

$$K_I = I_{1e} / I_{2e} = 100/5$$

It is so called rated current. The transformer is available for permanent operation under this current, which will not get damaged by heat. When load current exceeds the rated current, it is called load. Where the transformer is under overload operation for a prolonged time, it may result in burnout of its winding or reduced service life of insulating material.

2. Accuracy class

As there exists certain error to the current transformer, accuracy class of transformer is to be defined as per allowable error to current transformer. Current transformers made in China fall into the following accuracy classes: 0.2s, 0.5s, 0.2s, 0.5s, 5 and 10. 0.2s current transformer is mainly used for precise measurement at laboratory, or taken as a criterion for inspection of transformer of low classes. It can also be used in combination with any standard instrument for inspection of instruments. Therefore, it is also called standard current transformer. User electric energy measuring device normally uses Class 0.2 and Class 0.5 current transformer. It is applicable to select Class 5 and 10 current transformers for protection and control.

3. Rated Capacity

Rated capacity of current transformer refers to the apparent power S_{2e} consumed by the rated secondary current I_{2e} passing through the secondary rated load Z_{2e}. In this sense,

$$S_{2e} = I_{2e}^2 \times Z_{2e}$$

I_{2e} is normally = 5 A. Therefore, rated capacity can also be expressed in rated load impedance Z_{2e}.

Accuracy of current transformer in use can be ensured only on condition that total impedance of secondary wiring and instrument current coil is not over the rated capacity as indicated on the nameplate, and is no less than 1/4 rated capacity. Rated secondary load marked on manufacturer nameplate is normally expressed in rated capacity, of which standard output values include 5 V·A, 10 V·A, 15 V·A, 25 V·A, 30 V·A, 50 V·A, 100 V·A and so on.

4. Rated Voltage

Rated voltage for current transformer refers to the maximum voltage (valid value) that can be endured by primary winding for permanent grounding. It is only expected to indicate insulation intensity of current transformer, which has nothing to do with the rated capacity of current transformer at all. It is marked next to the model of current transformer. Taking LCW-35 for instance, "35" refers to the rated voltage that is in the unit of kV.

5. Polarity mark

To ensure accuracy of wiring for measurement and verification, polarity mark is to be provided on the terminal of primary and secondary winding for current transformer.

(1) Initial and terminal ends of primary winding are marked as L_1 and L_2 respectively. Where multiple limit primary winding is provided with taps, initial end is to be marked as L_1 followed in a proper sequence by L_2, L_3…from the first tap.

(2) Initial and terminal ends of secondary winding are marked as k_1 and k_2 respectively. Where the secondary winding is provided with intermediate taps, initial end is to be marked

as k_1, which is followed in proper sequence by k_2, k_3…from the first tap.

(3) As to current transformer with numerous secondary winding, it is necessary to further provide numbers before the mark "k" on the outgoing terminal of respective secondary winding, such as $1k_1$, $1k_2$, $1k_3$…; $2k_1$, $2k_2$, $2k_3$…

(4) The arrangement of sign symbols is requested to make sure that the secondary current flows out from k_1, and returns to k_2 via the external circuit when the primary current flow from end L_1 to end L_2.

Viewing from homopolar terminal for the primary winding and secondary terminal of current transformer, direction of current I_1 and I_2 is just opposite. Such a polar relationship is called subtractive polarity; on the contrary, it is called additive polarity. Current transformer is normally expressed in subtractive polarity.

3.3.2 Current Transformer Structure and Working Principles

3.3.2.1 Current transformer structure

Presently, current transformer used in power system is normally of electromagnetic type, of which basic structure is similar to that of common transformer. It comprises two mutually insulated winding (primary winding and secondary winding) as wound on the enclosed iron core, of which number of winding is represented by N_1 and N_2 respectively. See Fig. 3-16. The primary winding is in series connection with the circuit tested, and the secondary winding is in series connection with current coil of various measuring instruments or relays.

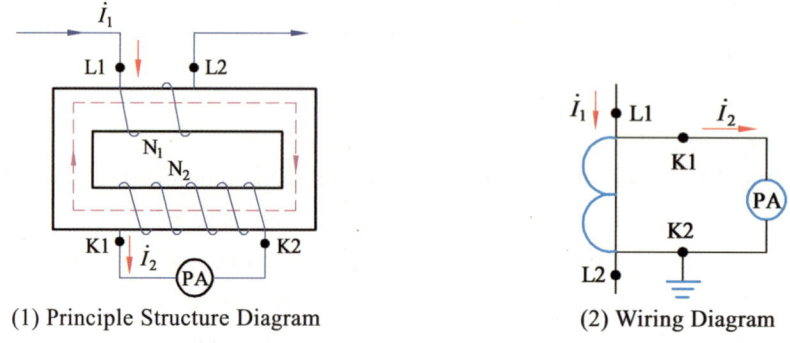

(1) Principle Structure Diagram   (2) Wiring Diagram

Fig. 3-16 Structural Diagram and Wiring Diagram for Current Transformer

In the power system, high current I_1 is frequently changed to low current I_2. Therefore, the number of turns N_2 for the secondary winding is over the number of turns N_1 of the primary winding. The secondary rated current for current transformer is normally 5 A or 1 A and 0.5 A. In the electrical diagram, current transformer is represented by Chinese symbol TA.

3.3.2.2 Working principles and characteristics

Working principles of current transformer are basically identical to that of common transformer. (When current $\dot{I}_1$ passes through the primary winding, most of fluxes produced by magnetomotive force $\dot{I}_1 N_1$ are to be closed via the iron core to induce electromotive force $\dot{E}_2$ in the secondary winding.) Where the secondary winding is connected with a load, current $\dot{I}_2$ will pass through the secondary winding. As current is accompanied by magnetomotive force, magnetomotive force $I_2 N_2$ will produce fluxes in the secondary winding. Most of fluxes are also enclosed via the iron core. On this account, fluxes in the iron core belong to synthetic fluxes Φ jointly produced by the primary winding and the secondary winding, which are called main fluxes. Based on magnetomotive force balance principle, it can be obtained that

$$\dot{I}_1 N_1 + \dot{I}_2 N_2 = \dot{I}_{10} N_1$$

In the formula, $\dot{I}_{10} N_1$ —excitation magnetomotive force.

Where various losses to iron core can be neglected, it can be considered that $\dot{I}_{10} N \approx 0$, then

$$\dot{I}_1 N_1 + \dot{I}_2 N_2 = 0$$
$$\dot{I}_1 N_1 = -\dot{I}_2 N_2$$

It is an extremely important comparison expression for an ideal current transformer. In other words, ampere-turn of primary magnetomotive force is equal to that of the secondary magnetomotive force, and the phase is just opposite. By further simplifying aforesaid expression, it can be obtained that

$$K_I = \frac{I_{1e}}{I_{2e}} = \frac{N_2}{N_1}$$

In other words, the rated current on both sides of an ideal current transformer is inversely proportional to their winding turns, and is equal to constant K_I, namely rated transformation ratio of current transformer.

Basic working principles and structural form of current transformer are similar to that of common transformer. However, working state of current transformers is significantly different from that of ordinary transformers:

(1) The primary current (I_1) of current transformer depends on the voltage and impedance of the primary circuit, which has nothing to do with the secondary load of current transformer. In other words, when the secondary load varies, for instance, increase or decrease of ammeters in series connection is unable to change its primary current value.

(2) The power consumed by the secondary circuit of current transformer will increase accompanied by increase in impedance to the secondary circuit, namely

$$S_2 = I_2^2 \times Z_2$$

(3) As load impedance to current transformer is mainly attributed to some instruments of extremely low internal resistance, such as current coil of ammeter and electricity meter, its working state is proximate to short-circuit state.

The iron core of ordinary current transformer is normally fabricated into a chip, which is made of high-quality silicon steel sheet. To minimize eddy current loss, adjacent two sheets are mutually insulated. Iron core of higher accuracy class for current transformer applied to the laboratory is made of permalloy, of which cross-section is in circular form. Such alloy features in higher initial permeability and extremely low loss.

3.3.3 Wiring Mode of Current Transformer

1. Two-phase star (V) connection

It comprises two current transformers. One end of the secondary winding of current transformer in connection with phase A and C is connected to the meter, and the other end is connected to phase B meter or the connection point at the outgoing terminal of phase a or c meter following mutual connection. Current to the secondary winding of the two current transformers is $\dot{I}_a$ and $\dot{I}_c$ respectively. Current passing through the public wiring is $\dot{I}_b = -(\dot{I}_a + \dot{I}_c)$. See Fig. 3-17. Such connection mode is normally applicable to three-phase-three-wire circuit. It has the following advantages:

(1) It can save conducting wires.

(2) It can obtain the 3rd phase current, normally phase B current, through wiring.

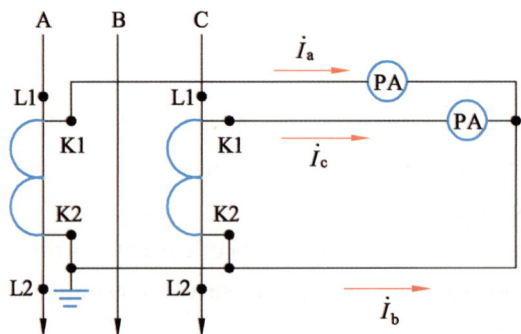

Fig. 3-17 Wiring Diagram for Two-phase Star (V) Principle

Nevertheless, such connection mode also has its disadvantages:

(1) As actual secondary load during verification with single phase method on site is inconsistent with that in operation, it is sometimes essential to use the three-phase method (or other similar methods). This may bring some difficulties to verification task.

(2) As it is probable that one of phases is reversely connected, common line current may become differential current to the extent of resulting in increase probability for wrong wiring. On this account, phase splitting method is used to electric energy metering circuit in some regions.

2. Phase splitting connection

Phase splitting connection refers to the fact that each phase is separately connected. See Fig. 3-18. It has the following advantages:

(1) The load during in-situ verification is identical to that during practical operation.
(2) The probability for wrong wiring is relatively low.

The only disadvantage is that it requires one additional conducting wire.

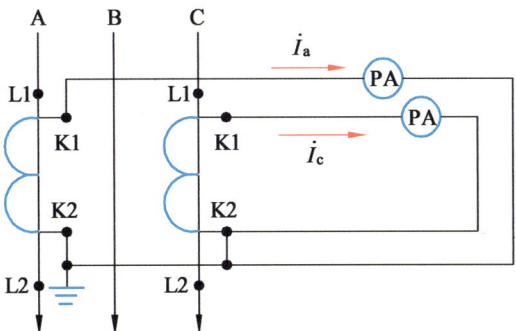

Figure 3-18 Wiring Diagram for Phase Splitting Principle

3. Three-phase Star (Y) Connection

Three-phase star connection is applied to most of three-phase-four-wire circuits. See Fig. 3-19. In the figure, current $\dot{I}_a$, $\dot{I}_b$, $\dot{I}_c$ andpasses through the secondary winding of current transformers at phase A, B and C respectively. When three-phase current is unbalanced, it is current $\dot{I}_N = \dot{I}_a + \dot{I}_b + \dot{I}_c$ to public wiring; it is current $\dot{I}_N = 0$ when three-phase current is balanced. Such wiring method allows no disconnection of public wiring. Otherwise, metering accuracy is to be affected (as there is no path for zero-sequence current).

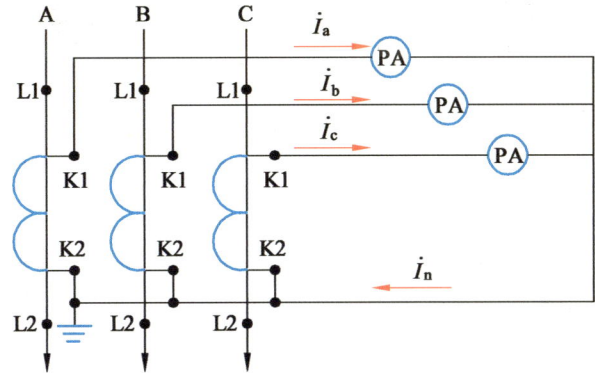

Fig. 3-19 Wiring Diagram for Three-phase Star (Y) Principle

3.3.4 Current Transformers Frequently Used

1. LFC-10 multi-turn wall penetration current transformer

As shown in Fig. 3-20 and 3-21, its primary winding penetrates through porcelain insulated bushing, and is fixed to flange plate; both ends are attached with a junction box; terminal block at lead-out of the primary winding is connected to the bus of power distribution unit. The secondary winding is installed inside enclosed housing, which is led out from wiring terminal.

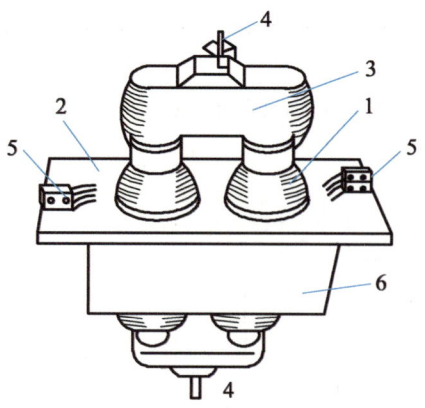 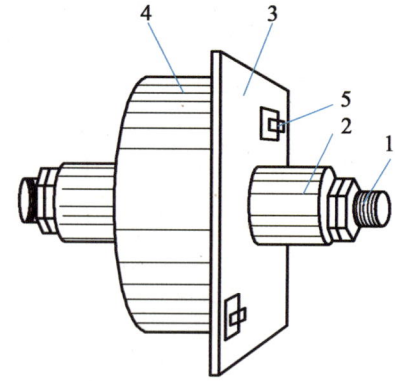

1—Porcelain insulated bushing 2—Flange plate; 3—Junction box; 4—terminal block ; 5—Wiring terminal of the secondary winding; 6—housing.

1—Primary winding; 2—Porcelain bushing; 3—Flange plate; 4—housing; 5—Wiring terminal of the secondary winding.

Fig. 3-20　LFC-10 Current Transformer 1　　　Fig. 3-21　LFC-10 Current Transformer 2

2. LDC-10 single-turn wall penetration current transformer

The series of current transformers are as shown in Fig. 3-22 and 3-23. The primary winding is a copper bar penetrating through the porcelain piece. The primary winding and porcelain piece penetrate into the core, and are fixed to flange plate. The secondary winding is installed in enclosed housing, which is led out from leading-out terminal.

Fig. 3-22　Physical Picture of LDC-10 Current Transformer 1　　　Fig. 3-23　Physical Picture of LDC-10 Current Transformer 2

3. LQJ-10 epoxy resin cast current transformer

This current transformer is as shown in Fig. 3-24. The primary and secondary winding is cast with epoxy resin. The primary winding is led out from leading-out terminal, and the secondary winding is led out from wiring terminal. Its major advantages include small volume, light weight and excellent electrical insulation. Presently, it is applied to power distribution unit of 10 kV or below. Physical picture of LQJ-10 current transformer is shown in Fig. 3-25.

4. LMC-10 bus type wall penetration current transformer

Its primary winding is the bus penetrating through porcelain bushing that penetrates into the core. Porcelain cap and clamp plate are provided on both ends of the porcelain bushing to clamp the bus. See Fig. 3-26 and 3-27.

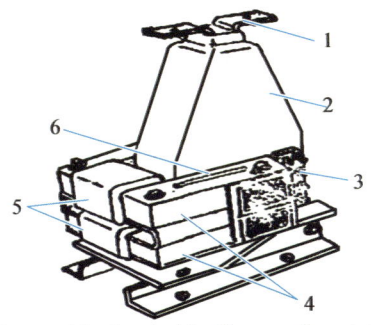

1—Primary wiring terminal 2—Epoxy resin cast primary winding; 3—Secondary wiring terminal 4—Iron core ; 5—Secondary winding; 6—Warning sign ("No Open Circuit to Secondary Winding")

Fig. 3-24　LQJ-10 Current Transformer

Fig. 3-25　Physical Picture of LQJ-10 Current Transformer

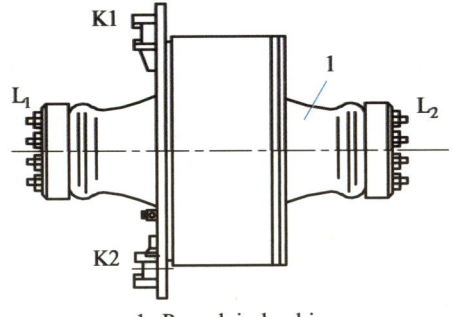

1- Porcelain bushin

Fig. 3-26　LMC-10 Current Transformer

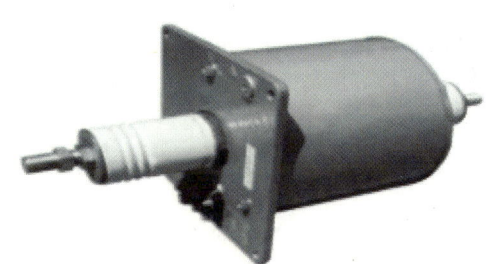

Fig. 3-27　Physical Picture of LMC-10 Current Transformer

任务一　互感器极性的测试方法
直流法判断互感器的极性

一、工作任务

按照规范的作业流程以及工艺要求，进行电流互感器、电压互感器的极性测试。掌握直流法进行互感器极性的判断的方法、工前准备、危险点预控、作业步骤、工艺要求及质量标准等操作技能。

二、引用标准

（1）《互感器试验导则》（GB/T 22071.1—2018）。
（2）《电流互感器试验导则》（JB-T 5356—2019）。
（3）《互感器运行检修导则》（DL/T 727—2013）。

三、工作要求

（1）作业为室外停电作业，避免雷雨及大风天气的影响。
（2）被检修设备与其他带电设备均应使用围栏隔离，面向通道处设置唯一出入口。
（3）被检修间隔可能来电端均停电并且接地。
（4）作业人员精神状态良好，熟悉工作中安全措施、技术措施以及现场工作危险点。
（5）实训现场要求按生产现场规范布置安全措施，并严格执行标准化作业。
（6）作业人员应规范穿戴劳动保护用品，做好安全防护。

四、工作准备

（一）危险点及预控措施

1. 高压触电

危险点：互感器一二次侧有可能突然来电，造成高压触电的危险。

预控措施：用围栏将被检修间隔与相邻带电设备（间隔）隔离，并且向作业现场内悬挂适量"止步，高压危险"标识牌，在通道处设置唯一出入口，悬挂"从此进出"标识牌；被检修间隔互感器可能来电侧装设短路接地线；工作时至少需要两人：一人监护，一人操作，听工作负责指挥。

2. 高处坠落

危险点：110 kV互感器多安装在户外变电站支架上，距离地面高度较高，人员在进行检修时必须注意防止高处坠落以及高空坠物。

预控措施：工作时至少需要两人，一人监护，一人操作，听工作负责指挥。人员在支架上工作时必须系安全带携带工具包，安全带高挂低用，严谨上下抛掷工具。

（二）工器具及材料选择

直流法判断互感器的极性工器具及材料见表3-3。

表3-3　直流法判断互感器的极性工器具及材料

类别	名称	规格型号	数量	备注
通用工具	电池		1.5~3 V	
	刀闸		1把	
	万用表	指针式	1只	
	活动扳手		1把	
	十字螺丝刀	端子螺丝刀	1把	
	连接导线	红黑	若干	
	放电棒		1根	
	绝缘手套		1双	
材料	砂纸		若干	
	抹布		若干	

（三）作业人员分工

直流法判断互感器的极性测试人员分工见表 3-4。

表 3-4　直流法判断互感器的极性测试人员分工

序号	工作岗位	数量	职责
1	工作负责人	1	负责本次工作的人员分工、现场查勘、作业方案制定、召开班前会、作业过程中安全监督、工作中突发状况的处理、工作质量的监督、班后会总结
2	操作人员	1	负责直流法判断互感器的极性测试的主要操作

五、工作程序

直流法判断互感器的极性测试作业流程见表 3-5。

表 3-5　直流法判断互感器的极性测试作业流程

序号	作业内容	作业步骤及标准	安全措施及注意事项
1	工作前准备工作	1. 检查工器具是否齐全，检查工器具外观和试验合格； 2. 工作负责人同工作许可人巡视待检修设备，确认工作票所列安全措施已经正确执行，安全措施是否完备，现场是否具备开工条件，必要时进行补充； 3. 执行工作许可手续； 4. 对工作班成员召开班前会； 5. 抄写设备铭牌参数	1. 工器具无损伤、变形、失灵现象，需要试验的工器具合格证在有效期内； 2. 巡视现场时禁止无关人员进入现场； 3. 班前会应包含工作地点双重名称；工作时间与内容；工作分工；工作危险点及预控措施；停电范围及工作现场安全措施； 4. 全体工作成员应当正确穿戴安全帽、工作服、工作鞋、劳保手套等劳动保护用品
2	确认设备状态	负责人带领工作班成员确认设备处于检修状态	这时检查（电流、电压）互感器有无接地，（注意：不要碰触电流、电压互感器）
3	互感器导通检查	1. 戴绝缘手套，将放电棒的接地端夹在互感器的外壳接地上，依次用放电棒的顶端（带接地电阻）和直接接地端钮对电流器一次侧、二次侧桩头进行充分放电，再对电压互感器一次侧、二次侧进行放电。将放电棒放在一侧。（放电棒接地线夹仍然夹在接地不要取下，后面要用）；	注意高处坠落和高空坠物风险

项目三　互感器检修

续表

序号	作业内容	作业步骤及标准	安全措施及注意事项
3	互感器导通检查	2. 分别悬挂"止步，高压危险""从此进出""在此作"标识牌； 3. 打磨清扫互感器。取砂纸对互感器桩头进行打磨，然后取抹布对互感器进行清扫。万用表导通检查。先将万用表挡位拔至"Ω"挡*1K挡位，再检查万用表，静态调零，在表头正、极开路的情况下，用罗丝批旋调万用表调零旋钮，使指"∞"位；动态调零，在万用表短路状态下，旋调万用表下"Ω"旋钮调零，使万用表指针指向"0"位； 4. 互感器导通检查。万用表在"Ω*1K"挡，用正负极测试夹分别碰及电流互感的一次侧、二次侧桩头，万用表应显示导通，再碰及电压互感器的一次侧、二次侧桩头，万用表应显示导通，以上说明电流、电压互感器一、二侧无断路现象	注意高处坠落和高空坠物风险
4	进行互感器极性检查。	1. 将万用表拔至A挡和50μA挡位； 2. 将电池与刀闸连接； 3. 电流互感器先将电源红色引线夹在L1桩头上（正极），将黑色引线（负极）夹在L2桩头上。将万用表正极引线夹在k1桩头上负极夹在k2桩头上。合上电源刀闸，如果万用表指针向右偏转，说明互感器为"减极性"，向左偏转说明互感器为"加极性"。试验连续进行三次。电流互感器极性测量完成后，断开电源，在取下测量线前，先戴绝缘手套，拿放电棒依次对电流互感器一次侧、二次侧桩头进行放电，然后取下电源及万用表引线； 4. 电压互感器将电源红色引线夹在一次侧A桩头上（正极），将黑色引线（负极）夹在一次侧B桩头上。万用表正极引线夹在二次侧a1桩头上，负极夹在二次侧b1桩头上。合上电源刀闸，观察万用表指针偏向判断极性；戴绝缘手套，拿放电棒对互感器一次侧、二次侧进行充分放电。取下电源、万用表引线，将万用表开关关闭，挡位旋钮旋至关闭。放回后面的指定位置	1. 充分放点； 2. 戴绝缘手套； 3. 其余二次端子依次进行
5	现场恢复至初始状态	1. 恢复互感器一次侧接线； 2. 恢复互感器二次侧接线； 3. 填写电流互感器、电压互感器极性试验记录单	
6	工作终结	1. 清理作业现场，做到"工完料尽场地清"； 2. 召开班后会； 3. 终结工作票	严禁负责人前去终结工作票的同时清理场地

直流法判断互感器的极性
电流互感器、电压互感器极性试验记录单

互感器极性试验记录：

天气：　　　　　　　　单位：　　　　　　　　地点：

互感器名称	型号	条形编码	外表检查结果	极性测试结论
电压互感器				
电流互感器				

试验人：

试验日期：

Task 1　Transformer Polarity Testing Method Determination of Transformer Polarity with DC Method

3.1　Work Tasks

Proceed with testing of polarity of current transformer and voltage transformer as specified operation procedures and technical requirements. Command approaches for determination of transformer polarity with DC method and such operation skills as work preparation, hazard prevention and control, operation steps, technical requirements and quality standards.

3.2　References

(1) *Test Guide for Instrument Transformers* (GB/T 22071.1—2018).

(2) *Test Guide for Current Transformers* (JB-T 5356—2019).

(3) *Guideline of Operation and Maintenance for Current and Voltage Transformers* (DL/T 727—2013).

3.3　Work Requirements

(1) The operation refers to outdoor power-cut operation that is to be protected from thunderstorm and strong wind.

(2) Equipments under maintenance are to be isolated from other electrified equipments by a fence. The only exit is to be provided at the place oriented towards the passage.

(3) Incoming terminal may subject to power cut and grounding at maintenance interval.

(4) Operators are to be in good mental state, and are aware of safety measures, technical measures and hazards to site operation during operation.

(5) Take safety measures on practical training site as per regulations on production site, and strictly implement standard operation.

(6) Operators are requested to wear labor protection appliances to ensure safety protection.

3.4 Preparation for Work

3.4.1 Hazards and preventive and control measures

1. High-voltage electric shock

hazards: Abrupt power-on on primary and secondary sides of transformer may result in high-voltage electric shock.

Preventive and control measures: Use fence to isolate maintenance interval from adjacent live equipments, and provide such sign boards as "Stop! High Voltage, Danger!" on operation site. The only exit is to be set at the passage, which is to be provided with the sign board indicating "Entrance/Exit"; Provide short-circuit grounding wire on the side of transformer that may subject to power-on at maintenance interval; At least two persons are required during work, one for supervision and the other for operation under instructions of person in charge of work.

2. High falling

Hazard: 110 kV transformer is normally installed on the support of outdoor transformer substation that is relatively higher above the ground. Operators are requested to guard against high falling and falling objects during maintenance.

Preventive and control measures: At least two persons are required during work, one for supervision and the other for operation under instructions of person in charge of work. Any operator working on the support is requested to wear safety belt, and carry a toolkit. Safety belt is to be hung high for use at a lower point. Throwing tools up and down is strictly prohibited.

3.4.2 Work tools and material selection

For instruments and materials used for determination of transformer polarity with DC method, please refer to Table 3-3.

Table 3-3 Instruments and Materials for Determination of Transformer Polarity with DC Method

Category	Name	Specification and model	Quantity	Remarks
General tools	Battery		1.5—3 V	
	Knife switch		1	
	Multimeter	Pointer type	1	
	Monkey wrench		1	
	Cross screwdriver	Terminal screwdriver	1	
	Connecting wire	R/B	Numerous	
	Discharging rod		1	
	Insulating gloves		1 pair	
Material	Abrasive paper		Numerous	
	Rag		Numerous	

3.4.3 Division of labor among operators

Work division for personnel engaged in testing transformer polarity with DC method is as shown in Table 3-4.

Table 3-4 Work Division for Personnel Engaged in Testing Transformer Polarity with DC Method

S/N	Job	Quantity	Responsibilities
1	Person in charge of work	1	Be responsible for work division of working staffs, site survey, stipulation of operation scheme, pre-shift meeting, safety supervision during operation, handling of emergencies during work, supervision of work quality and summary of post-shift meeting
2	Operator	1	Be responsible for major operations concerning testing of transformer polarity with DC method

3.5 Working Procedures

Testing procedures for determination of transformer polarity with DC method are as shown in Table 3-5.

Table 3-5 Testing Procedures for Determination of Transformer Polarity with DC Method

S/N	Scope of work	Operational steps and standards	Safety measures and precautions
1	Preparations before work	1. Check if all instruments and tools are complete, and if their appearance and test are acceptable; 2. The person in charge of work and work permitter shall make an tour inspection for equipments under maintenance, and confirm all	1. Instruments and tools are free of damage, deformation and malfunction. Qualification certificates for instruments and tools to be tested are within the term of validity;
1	Preparations before work	safety measures as listed by work ticket have been properly implemented. It is also necessary to check if safety measures are complete, and if the site is provided with conditions for commencement of work, and make supplements as required; 3. Implement work permit procedures; 4. Call in toolbox meeting participated by members of work team; 5. Record parameters on the equipment nameplate	2. Prevent other persons from entering the site during tour inspection; 3. Toolbox meeting shall cover dual designations of work place; Working hours and contents; Work division; Working hazards as well as preventive and control measures Power-cut scope and safety measures on work site; 4. All work members shall properly wear such labor protection appliances as safety helmet, working clothes, working shoes and protective gloves
2	Confirm equipment status	The person in charge shall lead members of work team to confirm that the equipment is ready for maintenance	In this case, check if transformer (current and voltage) is grounded (caution: never touch current or voltage transformer)
3	Check conduction of the transformer	1. Wear insulating gloves, and fix grounding terminal of discharging rod to the ground terminal on the transformer housing; use the top of discharging rod (with grounding resistance) and direct grounding terminal button to charge and discharge terminals on the primary and secondary sides of converter in turn; after that, further proceed with discharging to the primary and secondary sides of the voltage transformer. Place the discharging rod on one side. (Keep earthing clamp of discharging rod fixed to the grounding terminal for use in the future.) 2. Provide such sign boards as "Stop! High Voltage, Danger!", "Entrance/Exit" and "Work Here".	Guard against such risks as high falling and falling objects

Continued

S/N	Scope of work	Operational steps and standards	Safety measures and precautions
3	Check conduction of the transformer	3. Polish and clean the transformer. Use sandpaper to polish the transformer terminal, and then use rag to clean the transformer. Check conduction of multimeter. Firstly, switch the multimeter to* 1K at "Ω" position. After that, check the multimeter for static zero adjustment. Use screwdriver to rotate and adjust zero adjustment knob on the multimeter when positive and negative terminals on the outfit are in open circuit to make sure that the pointer points to "∞" position; Dynamic zero adjustment: Rotate and adjust "Ω" knob at lower part of the multimeter for zero adjustment when the multimeter is at short circuit status, and make sure that its pointer points to "0" position. 4. Check conduction of the transformer. Switch over multimeter to 1K at "Ω* position, and use positive and negative pole test clamp to touch terminals on the primary and secondary sides of current transformer to make sure that multimeter displays conduction status. After that, further touch terminals on the primary and secondary sides of voltage transformer to make sure that multimeter displays conduction status. Aforesaid operations are expected to indicate that there is no disconnection to primary and secondary sides of both current transformer and voltage transformer	
4	Check transformer polarity	1. Switch over multimeter to 50 μA at position A. 2. Connect the battery to the knife switch 3. Current transformer.	1. Carry out full discharge; 2. Wear insulating gloves; 3. Perform in other secondary terminals in turn

项目三　互感器检修

Continued

S/N	Scope of work	Operational steps and standards	Safety measures and precautions
4	Check transformer polarity	Firstly, fix red lead clamp of power source to terminal L1 (positive) and black lead clamp (negative) to terminal L2. Fix the positive lead clamp of multimeter to terminal k1 and negative one to terminal k2. Switch on knife switch of the power source. If the pointer of multimeter deflects to the right, it indicates that the transformer is in "subtractive polarity". On the contrary, if the pointer of multimeter deflects to the left, it indicates that the transformer is in "additive polarity". Carry out test for three consecutive times. Cut off power supply upon completion of measurement of polarity of current transformer. Wear insulating gloves prior to removal of measurement line, and use discharging rod to discharge terminals on the primary and secondary sides of current transformer. After that, remove the power source and lead wire of multimeter. 4. Voltage transformer. Fix red lead wire of power source to terminal (positive) on the primary side and black lead wire (negative) to terminal B on the primary side. Fix positive lead wire of multimeter to terminal a1 and negative lead wire to terminal b1 on the secondary side. Switch on the knife switch of the power source, and observe deflection of multimeter pointer to determine the polarity; Wear insulating gloves, and use discharging rod to proceed with full discharging to the primary and secondary sides of the transformer. Remove the power supply and multimeter leads, turn the multimeter switch off, and turn the gear knob to off. Put them back in the designated place	

Table 3-5

S/N	Scope of work	Operational steps and standards	Safety measures and precautions
5	Restore the site to its initial state	1. Restore the primary side wiring of transformer; 2. Restore the secondary side wiring of transformer; 3. Fill in the *Current Transformer and Voltage Transformer Polarity Test Record Sheet*	
6	End of work	1. Clean up the work site, and make sure that elements and materials are fully utilized and the site is cleaned; 2. Convene a post-shift meeting; 3. Terminate the work ticket	It is strictly forbidden for the person in charge to clean up the site while terminating the work ticket

Current Transformer and Voltage Transformer Polarity Test Record Sheet

Transformer polarity test record:

Weather:　　　　　　　　Unit:　　　　　　　　Location:

Transformer name	Model	Bar code	Exterior inspection result	Polarity test conclusion
Voltage transformer				
Current transformer				

Tester: _____

Date of test: _____

任务二　电流互感器的变比改接

一、工作任务

按照规范的作业流程以及工艺要求，进行电流互感器变比的改接。掌握 110 kV 电流互感器一次侧变比改接的方法、工前准备、危险点预控、作业步骤、工艺要求及质量标准等操作技能。

二、引用标准

（1）《互感器试验导则》（GB/T 22071.1—2018）。
（2）《电流互感器试验导则》（JB/T 5356—2019）。
（3）《互感器运行检修导则》（DL/T 727—2013）。

三、工作要求

（1）作业为室外停电作业，避免雷雨及大风天气的影响。

(2)被检修设备与其他带电设备均应使用围栏隔离,面向通道处设置唯一出入口。
(3)被检修间隔可能来电端均停电并且接地;
(4)作业人员精神状态良好,熟悉工作中安全措施、技术措施以及现场工作危险点。
(5)实训现场要求按生产现场规范布置安全措施,并严格执行标准化作业。
(6)作业人员应规范穿戴劳动保护用品,做好安全防护。

四、工作准备

(一)危险点及预控措施

1. 高压触电

危险点:电流互感器一二次侧有可能突然来电,造成高压触电的危险。

预控措施:用围栏将被检修间隔与相邻带电设备(间隔)隔离,并且向作业现场内悬挂适量"止步,高压危险"标识牌,在通道处设置唯一出入口,悬挂"从此进出"标识牌;被检修间隔互感器可能来电侧装设短路接地线;工作时至少需要两人,一人监护,一人操作,听工作负责指挥。

2. 高处坠落

危险点:110 kV 互感器多安装在户外变电站支架上,距离地面高度较高,人员在进行检修时必须注意防止高处坠落以及高空坠物。

预控措施:工作时至少需要两人,一人监护,一人操作,听工作负责指挥。人员在支架上工作时必须系安全带携带工具包,安全带高挂低用,严谨上下抛掷工具,使用工具必须抓稳拿牢。

(二)工器具及材料选择

电流互感器的变比改接工器具及材料见表 3-6。

表 3-6 电流互感器的变比改接工器具及材料表

类别	名称	规格型号	数量	备注
通用工具	活动扳手		2 把	
	活动扳手	大号	1 把	
	导电连接片		1 片	
材料	砂纸		若干	
	抹布		若干	
	导电膏		1 瓶	

(三)作业人员分工

电流互感器的变比改接人员分工见表 3-7。

表 3-7 电流互感器的变比改接人员分工表

序号	工作岗位	数量/人	职责
1	工作负责人	1	负责本次工作的人员分工、现场查勘、作业方案制定、召开班前会、作业过程中安全监督、工作中突发状况的处理、工作质量的监督、班后会总结
2	操作人员	1	负责变比改接工作的主要操作

五、工作程序

电流互感器的变比改接作业流程见表 3-8。

表 3-8 电流互感器的变比改接作业流程

序号	作业内容	作业步骤及标准	安全措施及注意事项
1	工作前准备工作	1. 检查工器具是否齐全，检查工器具外观和试验合格； 2. 工作负责人同工作许可人巡视待检修设备，确认工作票所列安全措施已经正确执行，安全措施是否完备，现场是否具备开工条件，必要时进行补充； 3. 执行工作许可手续； 4. 对工作班成员召开班前会； 5. 抄写设备铭牌参数	1. 工器具无损伤、变形、失灵现象，需要试验的工器具合格证在有效期内； 2. 巡视现场时禁止无关人员进入现场； 3. 班前会应包含工作地点双重名称；工作时间与内容；工作分工；工作危险点及预控措施；停电范围及工作现场安全措施； 4. 全体工作成员应当正确穿戴安全帽、工作服、工作鞋、劳保手套等劳动保护用品
2	确认设备状态	负责人带领工作班成员确认设备处于检修状态	检查电流互感器有无接地（注意：不要碰触电流互感器）
3	电流互感器一次绕组	电流互感器工作原理是利用电磁感应原理来实现小电流检测大电流。电流互感器的具体结构和原理这里就不做介绍，我们只需要知道两侧电流反比于线圈匝数，$I_1:I_2=N_2:N_1$。改电流互感器变比，就是改变一次的线圈匝数 N_1，从而改变电流互感器的量程。电流互感器的一次线圈呈"U"型结构，由两个半圆铝管（或铜管）构成。通过这两个半圆线圈串联和并联，实现一次线圈匝数的改变，从而更改电流互感器 CT 的变比	注意高处坠落和高空坠物风险

续表

序号	作业内容	作业步骤及标准	安全措施及注意事项
3	电流互感器一次绕组	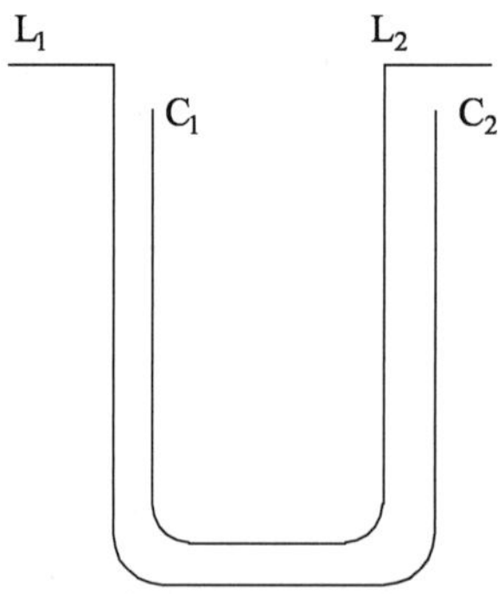 （图示：L_1、L_2、C_1、C_2 的U型线圈示意图）	
4	电流互感器一次侧串联绕组	当一次线圈串联时，线圈连接方式为：$L_1 \rightarrow C_2 \rightarrow C_1 \rightarrow L_2$，此时一次匝数为2匝（图示） 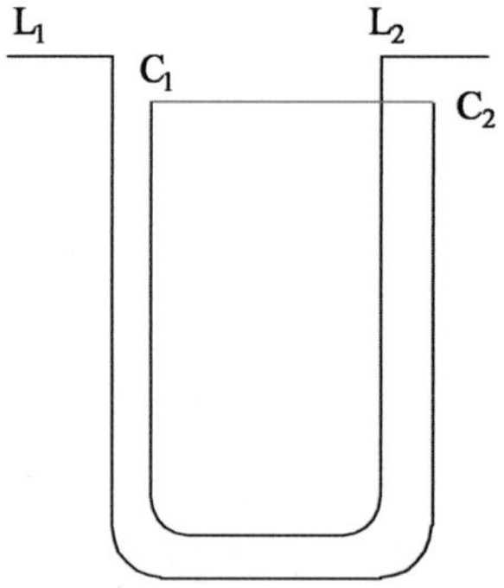	注意高处坠落和高空坠物风险
5	电流互感器一次侧并联绕组	当一次线圈并联时，线圈连接方式为：L_1、C_1相连，L_2、C_2相连，此时一次匝数为1匝（图示） 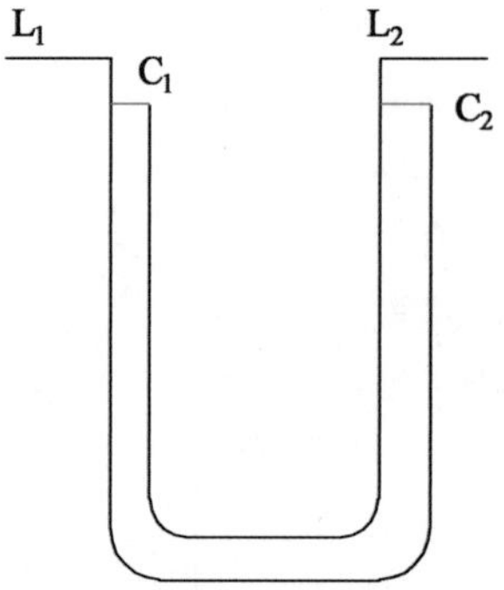	

续表

序号	作业内容	作业步骤及标准	安全措施及注意事项
6	现场恢复至初始状态	1. 恢复互感器一次侧接线； 2. 恢复互感器二次侧接线	
7	工作终结	1. 清理作业现场，做到"工完料尽场地清"； 2. 召开班后会； 3. 终结工作票	严禁负责人前去终结工作票的同时清理场地

Task 2　Wiring Change Based on Ratio of Transformation of Current Transformer

1. Work Tasks

Carry out wiring change based on ratio of transformation of current transformer in accordance with standardized operation procedures and process requirements. Master the method, pre-work preparation, hazard pre-control, operation steps, process requirements and quality standards and other operational skills of wiring change based on ratio of transformation of primary side of 110 kV current transformer.

2. References

(1) *Test Guide for Instrument Transformers* (GB/T 22071.1—2018).

(2) *Test Guide for Current Transformers* (JB-T 5356—2019).

(3) *Guideline of Operation and Maintenance for Current and Voltage Transformers* (DL/T 727—2013).

3. Work Requirements

(1) The operation refers to outdoor power-cut operation that is to be protected from thunderstorm and strong wind.

(2) Equipments under maintenance are to be isolated from other electrified equipments by a fence. The only exit is to be provided at the place oriented towards the passage.

(3) Incoming terminal may subject to power cut and grounding at maintenance interval.

(4) Operators are to be in good mental state, and are aware of safety measures, technical measures and hazards to site operation during operation.

(5) Take safety measures on practical training site as per regulations on production site, and strictly implement standard operation.

(6) Operators are requested to wear labor protection appliances to ensure safety protection.

4. Preparation for Work

(1) Hazards and preventive and control measures.

① High-voltage electric shock.

Hazard: The primary and secondary sides of the current transformer may be suddenly energized, resulting in the risk of high-voltage electric shock.

Preventive and control measures: Use fence to isolate maintenance interval from adjacent live equipments, and provide such sign boards as "Stop! High Voltage, Danger!" on operation site. The only exit is to be set at the passage, which is to be provided with the sign board indicating "Entrance/Exit"; Provide short-circuit grounding wire on the side of transformer that may subject to power-on at maintenance interval; At least two persons are required during work, one for supervision and the other for operation under instructions of person in charge of work.

② High falling.

Hazard: 110 kV transformer is normally installed on the support of outdoor transformer substation that is relatively higher above the ground. Operators are requested to guard against high falling and falling objects during maintenance.

Preventive and control measures: At least two persons are required during work, one for supervision and the other for operation under instructions of person in charge of work. Personnel working in the support must wear a safety belt and carry a toolkit. The safety belt shall be hung high and the personnel shall work below it. Throwing tools up and down is strictly prohibited, and tools must be grasped and held firmly.

(2) Work tools and material selection.

The tools and materials for wiring change based on ratio of transformation of current transformer are listed in Table 3-6.

Table 3-6　Tools and Materials for Wiring Change Based on Ratio of Transformation of Current Transformer

Category	Name	Specification and model	Quantity	Remarks
General tools	Monkey wrench		2	
	Monkey wrench	Large size	1	
	Conductive connector		1 Pcs	
Material	Abrasive paper		Numerous	
	Rag		Numerous	
	Conductive paste		1 Bottle	

(3) Division of labor among operators.

The work division of personnel for wiring change based on ratio of transformation of current transformer is listed in Table 3-7.

Table 3-7 Work Division of Personnel for Wiring Change Based on Ratio of Transformation of Current Transformer

S/N	Job	Quantity	Responsibilities
1	Person in charge of work	1	Be responsible for work division of working staffs, site survey, stipulation of operation scheme, pre-shift meeting, safety supervision during operation, handling of emergencies during work, supervision of work quality and summary of post-shift meeting
2	Operator	1	Be responsible for major operations concerning testing of transformer polarity with DC method

5. Working Procedures

Operation procedures for wiring change based on ratio of transformation of current transformer are listed in Table 3-8.

Table 3-8 Operation Procedures for Wiring Change Based on Ratio of Transformation of Current Transformer

S/N	Scope of work	Operational steps and standards	Safety measures and precautions
1	Preparations before work	1. Check if all instruments and tools are complete, and if their appearance and test are acceptable; 2. The person in charge of work and work permitter shall make an tour inspection for equipments under maintenance, and confirm all safety measures as listed by work ticket have been properly implemented. It is also necessary to check if safety measures are complete, and if the site is provided with conditions for commencement of work, and make supplements as required; 3. Implement work permit procedures; 4. Call in toolbox meeting participated by members of work team; 5. Record parameters on the equipment nameplate	1. Instruments and tools are free of damage, deformation and malfunction. Qualification certificates for instruments and tools to be tested are within the term of validity; 2. Prevent other persons from entering the site during tour inspection; 3. Toolbox meeting shall cover dual designations of work place; Working hours and contents; Work division; Working hazards as well as preventive and control measures Power-cut scope and safety measures on work site; 4. All work members shall properly wear such labor protection appliances as safety helmet, working clothes, working shoes and protective gloves

Continued

S/N	Scope of work	Operational steps and standards	Safety measures and precautions
2	Confirm equipment status	The person in charge shall lead members of work team to confirm that the equipment is ready for maintenance	Check the current transformer for grounding. (note: don't touch the current transformer)
3	Primary winding of current transformer	Current transformers work by utilizing the principle of electromagnetic induction to detect large currents with small currents. The specific structure and principle of the current transformer will not be introduced herein. We only need to know the current on both sides is inversely proportional to the number of turns of the coil, $I_1:I_2=N_2:N_1$. Changing the ratio of transformation of the current transformer is to change the number of turns of the primary coil N1, thus changing the range of the current transformer. The primary coil of the current transformer is in a "U" shape, consisting of two semicircular aluminum (or copper) tubes. Through the two semicircular coils in series and parallel, the number of turns of the primary coil is changed, and the ratio of transformation of the current transformer CT is changed thereof.	Guard against such risks as high falling and falling objects

Continued

S/N	Scope of work	Operational steps and standards	Safety measures and precautions
4	The current transformer is connected in series on the primary side	When the primary coil is connected in series, the coil is connected as follows: L1→C2→C1→L2 and the number of turns on the primary side is 2.	Guard against such risks as high falling and falling objects
5	The current transformer is connected in parallel on the primary side	When the primary coil is connected in parallel, the coil is connected as: L_1 is connected to C_1 and L_2 is connected to C_2, at which time the number of primary turns is 1.	
6	Restore the site to its initial state	1. Restore the primary side wiring of transformer; 2. Restore the secondary side wiring of transformer	
7	End of work	1. Clean up the work site, and make sure that elements and materials are fully utilized and the site is cleaned; 2. Convene a post-shift meeting; 3. Terminate the work ticket	It is strictly forbidden for the person in charge to clean up the site while terminating the work ticket

项目四　其他变电设备检修

模块一　母线

一、母线的作用

母线是将电气装置中各载流分支回路连接在一起的导体，起到汇集、分配、传送电能的作用。习惯上将各个配电单元中载流分支回路的导体均泛称为母线。

二、母线的材料

常用的母线材料有铜、铝和钢（见表 4-1）。

表 4-1　母线的材料

材料	优点	缺点	适用场合
铜	电阻率低，机械强度大，抗腐蚀性能强	储量少，价格高	持续工作电流较大的位置
铝	密度小，质量轻，储量丰富，价格低	抗腐蚀能力弱，机械强度低	屋内外配电装置
钢	机械强度高，价格低	电阻率大，集肤效应和磁滞、涡流损耗大	高压小容量电路、直流电路、电压互感器回路

铜的电阻率低、机械强度大、抗腐蚀性强，是很好的母线材料。但我国的铜储量不多，因此铜母线只在持续工作电流较大且位置特别狭窄的发电机、变压器出口处以及污秽对铝有严重腐蚀而对铜腐蚀较轻的场所（如靠近化工厂和海岸线）。

铝的电阻率为铜的 1.7~2 倍，但密度只有铜的 30%，在相同负荷及同一发热温度下，所耗铝的质量仅为铜的 40%~50%，而且我国的铝储量丰富，价格低，因此铝母线常用于屋内、外配电装置。但铝自身的机械强度较低，在常温下表面会迅速生成一层电阻率很大的氧化铝薄膜，且不易清除，此外铝的抗腐蚀性能较差，在铝、铜连接时会形成电位差，当接触面之间渗入含有溶解盐的水分时，可生成引起点解反应的局部电流，铝会被强烈腐蚀，使接触电阻增大，造成运行中温度升高，高温下腐蚀会更快，这样恶性循环导致接触处温度更高。所以在铜铝连接时，需要采用铜铝过渡接头，或在铜铝的接触表面搪锡。

钢的电阻率很大，为铜的 6~8 倍，用于交流时会产生很强的集肤效应，并造成很大的磁滞损耗和涡流损耗，但钢的机械强度高且价格低，因此仅用在高压小容量电路中（如电压互感器回路、小容量厂用、所用的变压器高压侧）、工作电流不大的直流电路或接地电路中。

三、母线的分类

按照外形和结构分,常用的母线分为硬母线、软母线、封闭母线三种。

所谓硬母线(见图4-1),是指用管型铝或者铜构成的导线,有矩形母线和管型母线之分,在外观上,硬母线像一块长条木板,主要应用在主变压器至配电室内,运行中变化小,载流量大,但造价较高。

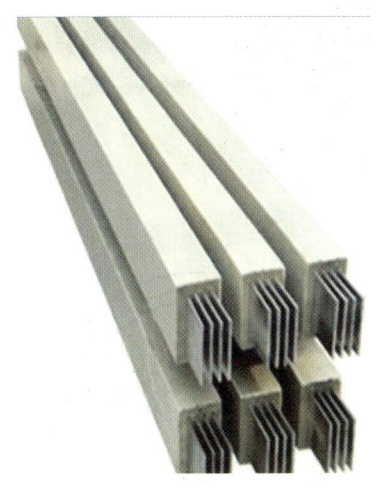 

图 4-1 硬母线

所谓软母线(见图4-2),是指用钢芯铝绞线或其他材料构成的导线,在外观上具有一定的柔软度,容易弯曲。软母线常用于室外,因室外空间大,导线即使有所摆动也不至于造成线间距不够。软母线施工简单,造价低廉。

 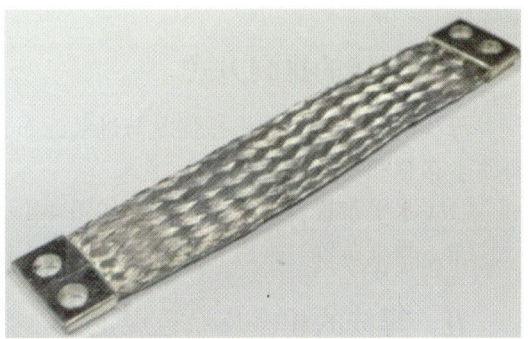

图 4-2 软母线

所谓封闭母线(见图4-3),是指由载流导体、壳体、以及绝缘材料组成的母线。母线封闭在壳体中,外壳接地,工作人员不会触及带电体,同时能有效防止绝缘遭受灰尘、潮气和外物导致的短路,因此封闭母线供电更加可靠安全,实际运行使用中维护工作量较小。封闭母线又分为共箱式封闭母线和分相封闭母线,共箱式封闭母线是指三相母线分别装设在支柱绝缘子上,共用一个金属薄板支撑的箱罩保护,三相母线之间可设金属隔板也可不设金属隔板；分相封闭母线的导体用支柱绝缘子支撑，一般有单个、两个、三个、四个绝缘子 4 种方案，国内设计的封闭母线几乎采用三个绝缘子方案，绝缘子在空间上互差 120°。

 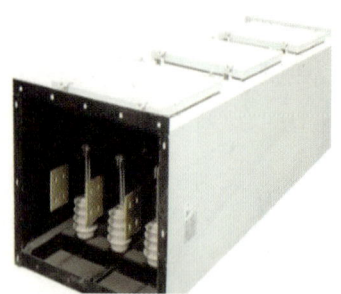

（a）共箱式封闭母线

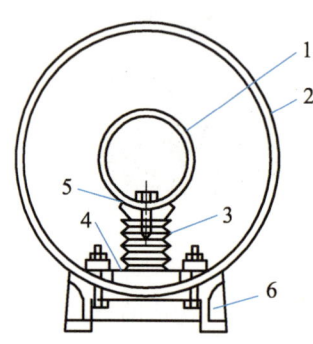 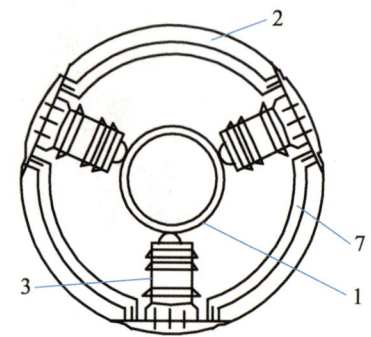

（b）分相封闭母线

1—载流导体；2—保护外壳；3—支柱绝缘子；4—弹性板；5—垫圈；6—底座；7—加强圈。

图 4-3　封闭母线

四、母线的排列方式

母线的散热条件和机械强度与母线的布置方式有关。最为常见的布置方式有两种，即水平布置和垂直布置。

（1）水平布置。水平布置方式如图 4-4（a）、（b）所示。

（2）垂直布置。垂直布置方式如图 4-4（c）所示。

五、母线的相序及涂色规定

（一）母线的相序排列

母线的相序排列(观察者从设备正面所见)原则如下：
从左到右排列时，左侧为 A 相，中间为 B 相，右侧为 C 相。
从上到下排列时，上侧为 A 相，中间为 B 相，下侧为 C 相。
从远至近排列时，远为 A 相，中间为 B 相，近为 C 相。
母线的相序和安装位置见表 4-2。

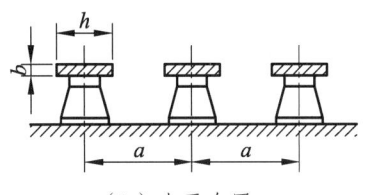

（b）水平布置

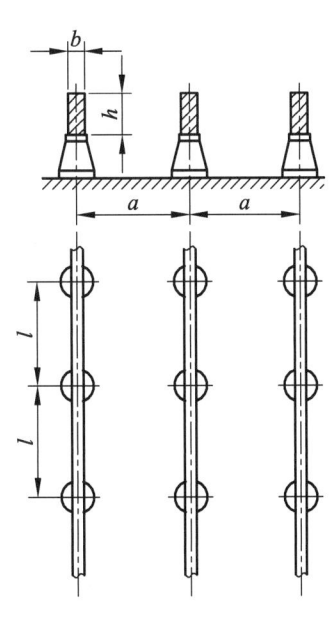

（a）水平布置

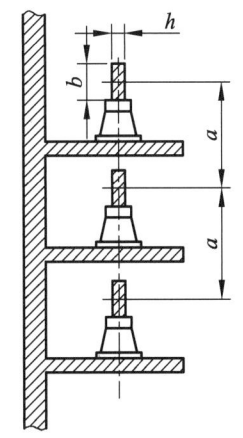

（c）垂直布置

图 4-4 母线的布置方式

表 4-2 母线的相序和安装位置

相别	颜色	母线安装相互位置		
		垂直排列	水平排列	引下线
A 相	黄	上	远	左
B 相	绿	中	中	中
C 相	红	下	近	右

（二）母线的涂漆颜色规定

涂色：A—黄色，B—绿色，C—红色，接地线—黑色。

母线相序为：A，B，C 标色为：黄，绿，红。

安装排列次序：

由上至下：黄，绿，红。

由左至右：黄，绿，红。

由后至前：黄，绿，红。

Program 4 Maintenance of Other Transformation Equipment

Module 1 Bus

4.1.1 Role of Bus

Bus is a conductor that connects the current carrying branch circuits of an electrical equipment together and is used for gathering, distributing and transmitting electric energy. Conventionally, the conductors of the current carrying branch circuits in each distribution unit are generally referred to as bus.

4.1.2 Bus Material

Commonly used bus materials are copper, aluminum and steel (see Table 4-1).

Table 4-1 Bus Material

Material	Advantages	Disadvantages	Occasion applicable
Copper	Low resistivity, high mechanical strength and corrosion resistance	Low reserves and high price	Positions with high continuous working current
Aluminum	Low density, light weight, abundant reserves and low price	Low corrosion resistance and mechanical strength	Indoor and outdoor power distribution units
Steel	High mechanical strength and low price	High resistivity, high skin effect and hysteresis, eddy current losses	High-voltage and small-capacity circuits, DC circuits, voltage transformer circuits

Copper's low resistivity, high mechanical strength and corrosion resistance make it a good bus material choice. However, China's copper reserves are not large, so the copper bus is only used in generators and transformer exits where the continuous working current is large and the space is particularly narrow, and places where the pollution has serious corrosion on aluminum and light corrosion on copper (such as places near the chemical plant and the coastline).

Aluminum resistivity is 1.7—2 times that of copper, but aluminum density is only 30% of that of copper. In the same load and the same heating temperature, the weight of aluminum consumed is only 40%—50% of that of copper. Moreover, China's aluminum reserves are abundant, the price is low, so the aluminum bus is commonly used in the indoor and outdoor power distribution units. However, the mechanical strength of aluminum is low, and a layer of alumina film with high resistivity will be quickly formed on the surface at room temperature, and it is not easy to be removed. In addition, the corrosion resistance of aluminum is poor. When aluminum and copper are connected, a potential difference will be formed, and when water containing dissolved salt is infiltrated between the contact surfaces, local currents that cause electrolytic reactions can be generated, and the aluminum will be strongly corroded, resulting in higher contact resistance and higher temperature in operation. Corrosion will be faster at high temperatures, resulting in a vicious circle that leads to higher temperatures at the contact. Therefore, when connecting copper and aluminum, it is necessary to use copper and aluminum transition joints, or perform tin lining on the contact surface of copper and aluminum.

The resistivity of steel is very large, which is 6 to 8 times that of copper. It will produce strong skin effect and cause great hysteresis loss and eddy current loss when used in AC, but the mechanical strength of steel is high and the price is low. Therefore, it is only used in high-voltage and small-capacity circuits (such as voltage transformer circuit, small-capacity plant, HV side of transformer), DC circuit or grounding circuit with low working current.

4.1.3 Classification of Bus

According to the shape and structure, the commonly used buses are divided into hard bus, soft bus and enclosed bus.

The hard bus (see Fig. 4-1) refers to a wire composed of tubular aluminum or copper, including rectangular bus and tubular bus. In appearance, the hard bus is like a long plank of wood, mainly used in the main transformer to the distribution room, with small changes in operation and large current carrying capacity, but the cost is high.

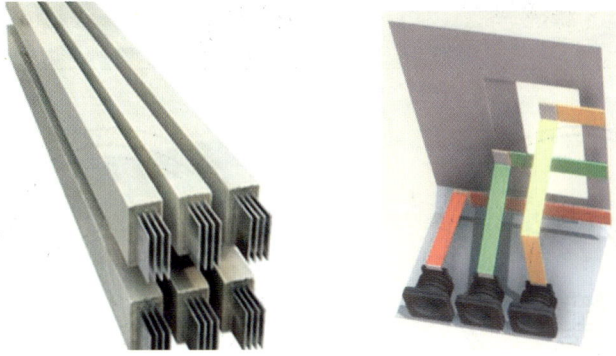

Fig. 4-1 Hard Bus

The soft bus (see Fig. 4-2) refers to a wire constructed of steel-core aluminum stranded wire or other materials, which is soft in appearance and easy to bend. Soft buses are often used outdoors, where there is plenty of space, and even if the wires swing, they do not cause insufficient spacing between the wires. Soft bus construction is simple and inexpensive.

 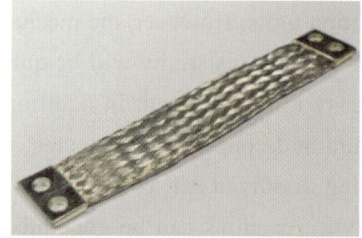

Fig. 4-2　Soft Bus

The enclosed bus (see Fig. 4-3) refers to a bus that consists of a current carrying conductor, a housing, and insulating materials. The bus is enclosed in the housing and the housing is grounded, so the staff will not touch the charged object, and at the same time, it can effectively prevent the insulator from short circuit caused by dust, moisture and external objects. Therefore, the power supply of the enclosed bus is more reliable and safe, and the maintenance workload is less in actual operation. Enclosed buses are also divided into common enclosure bus and split-phase enclosure bus. With regard to the common enclosure bus, the three-phase buses are installed in the post insulators respectively and protected by a box cover supported by a metal sheet, and the three-phase buses can be provided with or without a metal partition. The conductor of the split-phase enclosure bus is supported by post insulator, and there are generally four schemes of single insulator, two insulators, three insulators and four insulators. In the domestic design of the enclosed bus, the scheme of three insulators is often adopted, and the insulators have a spatial mutual deviation of 120°.

(a) Common enclosure bus

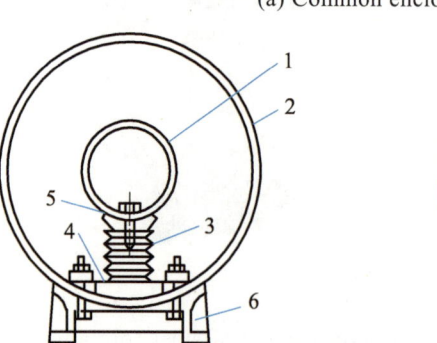 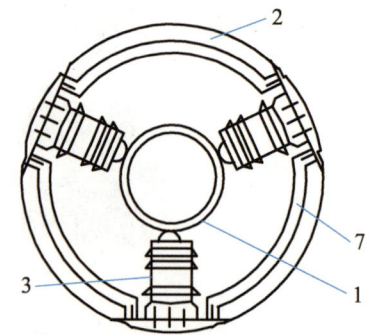

(b) Split-phase enclosure bus

1—Current carrying conductor; 2—Protective housing; 3—Post insulator; 4—Elastic plate;
5—Washer; 6—Base; 7—Reinforcement ring.

Fig. 4-3　Enclosed Bus

4.1.4 Arrangement of Bus

The heat dissipation condition and mechanical strength of the bus are related to the arrangement of bus. The two most common types of arrangement are horizontal and vertical.

(1) Horizontal arrangement. The horizontal arrangement is shown in Fig. 4-3 (a) and (b).

(2) Vertical arrangement. The vertical arrangement is shown in Fig. 4-3 (c).

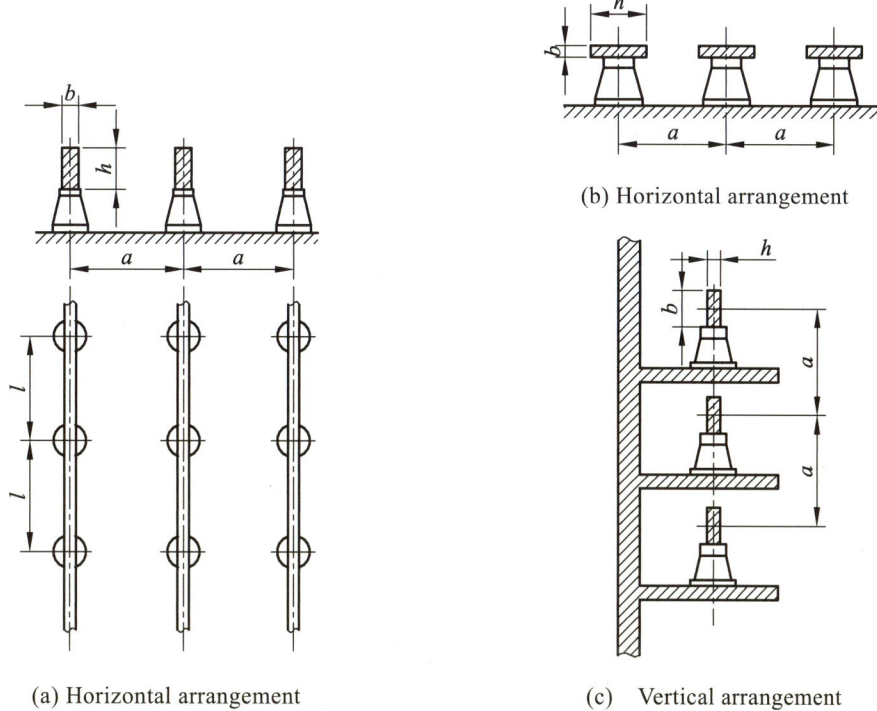

(a) Horizontal arrangement (c) Vertical arrangement

Fig. 4-4　Arrangement of Bus

4.1.5　Phase Sequence and Coloring Provisions for Bus

1. Phase sequence of bus

The phase sequence of bus (as seen by the observer from the front of the equipment) is based on the following principles:

When phases are arranged from left to right, phase A is on the left, phase B in the center, and phase C on the right.

When phases are arranged from top to bottom, Phase A is on the top side, Phase B in the center, and Phase C on the bottom side.

When phases are arranged from far to near, phase A is at the far end, phase B in the center, and phase C in the near end.

Phase sequence and installation location of bus see Table 4-2.

Table 4-2　Phase Sequence and Installation Location of Bus

Phase	Color	Installation location of bus		
		Vertical arrangement	Horizontal arrangement	Down conductor
Phase A	Yellow	Top side	Far end	Left
Phase B	Green	Center	Center	Center
Phase C	Red	Bottom side	Near end	Right

2. Coloring provisions for bus

Colors: A—yellow, B—green, C—red, grounding wire—black.

Phase sequence of bus: A, B, C. Colors: Yellow, green, red.

Sequence of installation:

From top to bottom: Yellow, green, red.

From left to right: Yellow, green, red.

From back to front: Yellow, green, red.

模块二　绝缘子

一、绝缘子的作用

绝缘子广泛应用在发电厂的配电装置、变压器、开关电器及输电线路上，用来支撑和固定裸载流导体，并使载流导体与地绝缘，或使处于不同电位的载流导体之间绝缘。

二、绝缘子的材料

绝缘子材料通常有陶瓷（见图 4-5）、玻璃和复合材料三种。高压绝缘子主要由绝缘件和金属附件两部分构成。

图 4-5　陶瓷绝缘子

瓷绝缘子的绝缘件由电工陶瓷制成，电工陶瓷由石英、长石和黏土做原理烘焙而成。瓷绝缘子的表面通常由瓷釉覆盖，用以提高其机械强度，防水浸润，增加表面的光滑度，在各类绝缘子中，瓷绝缘子使用最为广泛。

玻璃绝缘子（见图 4-6）的绝缘件由钢化玻璃制成，具有绝缘和机械强度高、尺寸小、质量轻、制造工艺简单及价格低廉等优点，且与瓷绝缘子不同，玻璃绝缘子具有零值自爆的绝缘自我淘汰能力，当丧失绝缘性能时容易被发现，故无需对其进行绝缘测试。

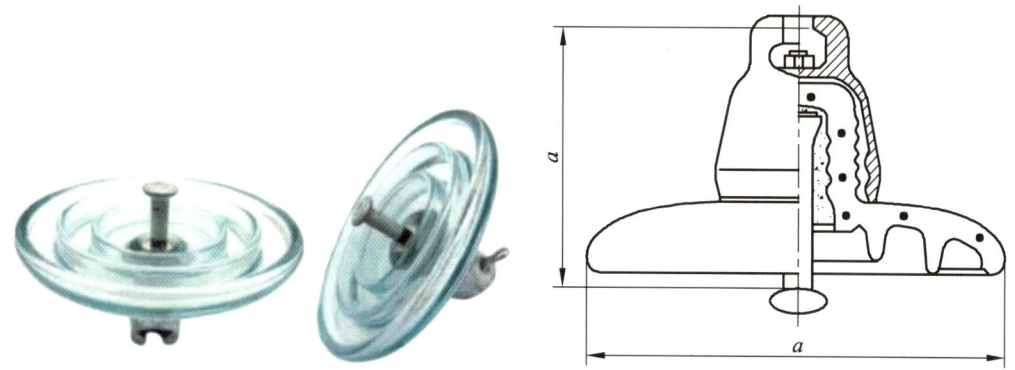

图 4-6　玻璃绝缘子

复合绝缘子（见图 4-7）一般是指硅橡胶绝缘子，它由伞盘、芯棒及金属端头三部分组成，其中伞盘由热硫化硅橡胶制造，芯棒采用环氧树脂经特殊工艺制成。同传统的陶瓷或玻璃绝缘子相比，具有许多优点，主要表现在：

（1）质量轻、强度高。由于其主要构成材料是容重比陶瓷和玻璃低得多的硅橡胶和环氧树脂，而且只有两端的金属联接附件，整体重量只及陶瓷和玻璃串绝缘子的 1/10～1/7，而其拉伸强度却高出 3～4 倍。

（2）电绝缘及防污闪性能优良。由于伞裙、护套均采用表面能很低、憎水性能好、抗紫外和抗臭氧等耐候性能优良，而且选择具有非延燃和自熄性能的高绝缘硅橡胶材料制成，大大改善了高压绝缘设备表面的受污、受潮状况，提高绝缘强度和耐污能力，有效地提高用电安全性，基本上不用人工清洁，减少了维护费用。

（3）整体结构增大了爬距比。由于硅橡胶高压复合绝缘子在结构上为整只构成，上下只有两个电极，因此其爬距比远大于陶瓷及玻璃绝缘子串，在某种程度上也增强其电绝缘性能。

（4）有利于城市架空电力线路的小型化。由于其质量轻，可以减轻杆塔荷重，使线路间距缩小，线路走廊宽度减小，显示出质量轻、效能高、总费用低的优点，有利于城市架空线路小型化，为高压输电线路提供了新的技术条件。

图 4-7　复合绝缘子

复合绝缘子优点虽多，但在长期运行中也难免出现各种问题和事故，如芯棒脆断、酥断、局部温升、界面局部粘黏不牢、老化、憎水性丧失等问题。

对绝缘子来说，影响其市场和应用通常有三个方面：价格、性能和维护成本。在价

项目四　其他变电设备检修　249

格方面，几乎在所有电压等级下复合绝缘子的价格均小于瓷和玻璃绝缘子，而且电压等级越高、吨位越大，价格差别越大，在直流情况下更为明显。在性能方面，主要是指耐污性能、电气性能、机械性能和长期性能。复合绝缘子的耐污性能远优于瓷和玻璃绝缘子，电气性能与其他两者相似，机械性能接近或稍差于其他两者，而在长期性能上瓷和玻璃绝缘子更有优势。在运行维护成本方面，复合绝缘子由于无零值、免清扫等特性，比其他两者更有优势。

电力部门在选择绝缘子时，上述三个方面的重要性排序会根据运行环境不同而有所差别。在我国，多数绝缘子运行在中等污秽或污秽程度更高的环境中，综合考虑线路造价、安全性等因素，三者的重要性顺序实际为价格、性能和维护成本。

可以预见，在今后若干年内，硅橡胶复合绝缘子由于低廉的价格，优异的耐污性能和维护简单的特点，仍将是电力部门的首选。

三、绝缘子的分类

1. 按额定电压分

高压绝缘子（500 V 以上）、低压绝缘子（500 V 以下）（见图 4-8）。

（a）高压玻璃绝缘子

（b）低压陶瓷绝缘子

图 4-8 高压玻璃绝缘子和低压陶瓷绝缘子

2. 按安装地点分

户内式和户外式（见图 4-9）。

（a）户内绝缘子

（b）户外绝缘子

图 4-9 户内绝缘子和户外绝缘子

3. 按结构形式和用途分

支柱式、套管式、盘形悬式。

（1）支柱式绝缘子。

支柱式绝缘子是一种特殊的绝缘控件，能够在架空输电线路中起到重要的作用。早年间支柱式绝缘子多用于电线杆，慢慢发展于高压电线连接塔的一端挂了很多悬状的绝缘体，它是为了增加爬电距离，通常由硅胶或陶瓷制成。绝缘子在架空输电线路中起着两个基本作用，即支撑导线和防止电流回地，这两个作用必须得到保证，绝缘子不应该产生由于环境和电负荷条件发生变化导致的闪络击穿而失效，否则绝缘子就会失去作用，会损害整条线路的使用和运行寿命。

支柱绝缘子均符合《高压支柱瓷绝缘子技术条件》（GB 8287.1）和《耐污型户外棒形支柱瓷绝缘子》（GB 12744）的规定，也符合国际标准 IEC《标称电压高于 1000 伏的系统用户内和户外瓷或玻璃支柱绝缘子的试验》及 IEC 出版物《绝缘子在污秽条件下的选用导则》的规定，具有以下特点：

① 机械强度高，分散性小，运行安全可靠。
② 低温性能好。
③ 耐污性能好。
④ 耐地震水平高。
⑤ 无线电干扰低。

（2）套管式绝缘子。

套管绝缘子用于母线在屋内穿过墙壁或天花板以及从屋内向屋外引出，或用于使有封闭外壳的电气（如断路器、变压器等）的载流部分引出壳外。套管绝缘子也称为穿墙套管，简称套管。

穿墙套管按安装地点可分为户内式和户外式两种，根据结构形式可分为带导体型和母线型两种。带导体型套管，其载流导体与绝缘部分制成一个整体，导体材料有铜和铝，导体截面有矩形的和圆形的；母线型套管本身不带载流导体，安装使用时，将载流母线安装于套管的窗口内。

（3）盘形悬式绝缘子（见图4-10）。

1—脚；2—帽。

图4-10 盘形悬式绝缘子

悬式绝缘子主要应用在 35 kV 及以上屋外配电装置和架空线路上，金具一般由钢冒和针形钢脚构成，故盘形悬式绝缘子又称为帽-脚形悬式绝缘串子，按其帽及脚的连接方式，分为球形和槽型两种。

悬式绝缘子均由绝缘件（瓷件或钢化玻璃）、铁帽、铁脚组成。钟罩形防污绝缘子的污闪电压比普通绝缘子高 20%～50%；双层伞形防污绝缘子具有泄露距离大、伞形开放、裙内光滑、积灰率低、自洁性能好等优点；草帽形防污绝缘子具有积污率低，自洁性能好等优点。

在实际应用中，悬式绝缘子根据装置电压的高低组成绝缘子串。这时，一片绝缘子的脚的粗头穿入另一片绝缘子的帽内，并用特制的弹簧锁锁住。不同电压等级每串绝缘子的数目不同，一般 35 kV 不少于 3 片，110 kV 不少于 7 片，220 kV 不少于 13 片，330 kV 不少于 19 片，500 kV 不少于 24 片，具体可根据当地污秽程度增加绝缘子片数，对于严重污秽处，应选用防污悬式绝缘子。

4. 按用途分

分为电站绝缘子、电器绝缘子、线路绝缘子。

（1）电站绝缘子：主要用来支持和固定发电厂及变电站屋内外配电装置的硬母线，并使母线与大地绝缘。按作用不同分为支柱绝缘子和套管绝缘子。

（2）电器绝缘子：主要用来固定电器的载流部分。也分为支柱绝缘子和套管绝缘子。支柱绝缘子用于固定没有封闭外壳的电器的载流部分；套管绝缘子用来使有封闭外壳的电器（如断路器、变压器等）的载流部分引出外壳。

（3）线路绝缘子：主要用来固结架空输、配电导线和屋外配电装置的软母线，并使它们与接地部分绝缘。有针式、悬式、蝴蝶式和瓷横担四种。

四、绝缘子故障及其防止措施

绝缘子是配电线路上的重要元件之一，靠它把导线与大地绝缘，一旦绝缘子发生损坏，带电导线对地失去绝缘，将造成接地短路或相间短路故障，从而造成线路供电中断。

1. 绝缘子因制造不良而损坏的故障

良好合格的绝缘子，瓷质细腻洁白，表面涂釉均匀，进行各种电气试验和温差实验时均能达到合格标准。电气试验包括：干闪试验、湿闪试验、耐压试验、温差试验等，质量较差的绝缘子无法通过上述试验。近年来，农村中架设的配电线路上采用了瓷质粗劣、性能较差的针式绝缘子和悬式绝缘子，线路架成投入运行后不久或经过雨季便发生闪络、击穿和炸裂等事故。

劣质瓷质针式绝缘子和悬式绝缘子共同的特点是雨天吸潮、绝缘迅速降低，从而导致闪络和击穿。将劣质绝缘子打碎后，其瓷质粗糙、略呈褐黄色，滴上墨水迅速扩散；有些针式绝缘子在胶合铁脚时采用了硫磺和氧化铅等作胶合剂，容易受热膨胀而使绝缘子炸裂。

防止措施：在线路架设安装前对绝缘子进行试验，采用厂家不明或非电气专业工厂生产的绝缘子时，应进行全面的电气试验，并将绝缘子浸水数小时后再做干闪、湿闪和耐压试验。同时还应抽取绝缘子做温差试验，凡浸水后试验不合格的绝缘子不能在线路上使用。

2. 施工不当使绝缘子损伤的故障

在架设线路过程中，由于运输或施工时的不慎，使绝缘子收到损伤，如裂纹、破损、瓷釉脱落等，裂纹和硬伤面积超过 100 mm² 以上的绝缘子，如用于线路上，投入运行后在阴雨天气就会发生击穿或闪络故障。

防止措施：使用前仔细观察检查绝缘子状况，避免使用裂纹和硬伤面积过大的绝缘子。

3. 外力损伤造成的绝缘子损坏

线路上的绝缘子有时会被抛掷的砖块、石子或弹弓弹射击中造成裂纹、硬伤、破损等。绝缘子发生裂纹或严重的硬伤后有时在晴朗天气还能继续运行，但遇小雨、重雾天气绝缘就急剧下降，以致发生闪络或绝缘击穿接地故障。

防止措施：发现缺陷绝缘子时应立即更换，但如果绝缘子硬伤较轻，且硬伤部位远离导线，则可在硬伤处涂以绝缘漆，不必更换。

4. 绝缘子老化造成的故障

正常运行中的绝缘子承受着高电压，也承受较大的拉力，经过长期运行，其绝缘会逐渐老化，绝缘降低，以致发生故障，严重老化了的绝缘子，其绝缘电阻已降低到不能继续运行的成都，需立即更换。

防止措施：运行人员应采用带电测试方法进行试验，发现老化绝缘子应及时移除，此外也可利用线路停电机会，使用绝缘摇表测绝缘子的绝缘电阻，通常运行中合格的绝缘子其最低绝缘电子应在 300 MΩ 以上。

Module 2　Insulator

4.2.1　Role of Insulator

Insulators are widely used in power distribution units, transformers, switching devices and transmission lines in power plants to support and secure bare current carrying conductors and to insulate the current carrying conductors from the ground or to insulate the current carrying conductors at different potentials from each other.

4.2.2　Insulator Material

There are usually three kinds of insulator materials: ceramics, glass and composite materials. High-voltage insulators are mainly composed of two parts: insulating parts and metal accessories.

The insulating parts of ceramic insulators are made of electrical ceramics, which are baked from quartz, feldspar and clay. The surface of the ceramic insulator is usually covered by enamel to improve its mechanical strength and waterproof infiltration as well as increase the smoothness of the surface. Among all types of insulators, ceramic insulators (see Fig.4-5) are most widely used.

Fig. 4-5　Ceramic Insulator

The insulating parts of glass insulators (see Fig.4-6) are made of tempered glass, which have the advantages of high insulation and mechanical strength, small size, light weight, simple manufacturing process and low price, and are different from the insulating parts of ceramic insulators. Glass insulators have the insulation self-elimination ability of zero self-explosion. When the insulation performance is lost, it is easy to be found, so it is not necessary to carry out insulation testing.

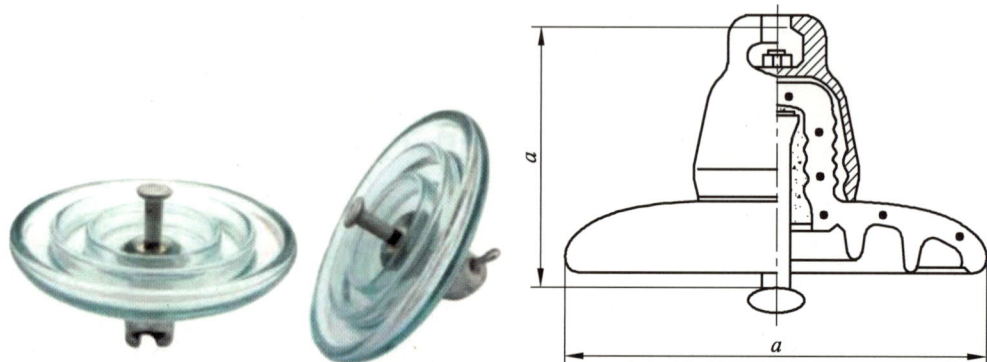

Fig. 4-6　Glass Insulator

Composite insulator (see Fig.4-7) generally refers to silicone rubber insulator, which consists of three parts: umbrella tray, mandrel and metal end, of which the umbrella tray is made of hot vulcanized silicone rubber, and the mandrel is made of epoxy resin through a special process. Compared with conventional ceramic or glass insulator, it has many advantages, mainly including:

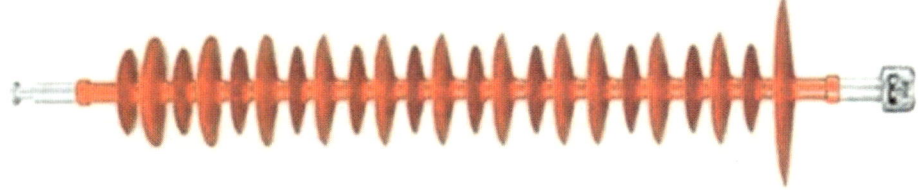

Fig. 4-7 Composite Insulator

(1) Light weight and high strength. Because its main composition material is silicone rubber and epoxy resin whose bulk density is much lower than that of ceramics and glass, and there are only metal connection accessories at both ends, the overall weight is only 1/10—1/7 of the ceramic and glass insulator strings, but its tensile strength is 3—4 times higher than that of the ceramic and glass string insulators.

(2) Excellent electrical insulation and anti-pollution flashover performance. Since the umbrella skirt and sheath are made of high insulating silicone rubber materials with very low surface energy, good hydrophobic performance, excellent ultraviolet and ozone and weather resistance, and non-delayed combustion and self-extinguishing performance, the pollution and moisture on the surface of high-voltage insulation equipment is greatly alleviated, the insulation strength and pollution resistance is enhanced, the safety of electricity consumption is effectively improved, no manual cleaning is needed, and the maintenance cost is reduced.

(3) As for the overall structure, the specific creepage distance is increased. Since high-voltage composite insulators are structurally made of silicone rubber as a whole, there are only two electrodes up and down, so its specific creepage distance is much larger than that of ceramic and glass insulator strings, and its electrical insulation performance is enhanced to some extent.

(4) Miniaturization of urban overhead power lines is facilitated. Because of its light weight, it can reduce the load of pole and tower, reduce the line spacing and reduce the line corridor width, showing the advantages of light weight, high efficiency and low total cost, which is conducive to the miniaturization of urban overhead lines, and provides new technical conditions for high-voltage transmission lines.

Although there are many advantages of composite insulator, there are also various problems and accidents in long-term operation, such as the brittle break, crisp break of mandrel, local temperature rise, weak local adhesion of interface, aging, loss of hydrophobicity and so on.

For insulators, there are typically three elements that affect their markets and applications: Price, performance and maintenance costs. In terms of price, the price of composite insulators is less than that of porcelain and glass insulators at almost all voltage classes, and the higher the voltage level is and the larger the tonnage is, the greater the price difference is, especially in the case of DC. In terms of performance, the main focus is pollution resistance, electrical performance, mechanical performance and long-term performance. Composite insulator is far better than ceramic and glass insulators in terms of pollution resistance, similar to the other two types of insulators in terms of electrical performance, close to or slightly worse than the other two types of insulators in terms of mechanical performance, and more advantageous than the other two types of insulators in terms of the long-term performance. With regard to operation and maintenance costs, composite insulator is more advantageous than the other two types of insulators due to its features such as no zero value and no cleaning.

When the power sector chooses insulators, the order of importance of the above three elements will vary according to the operating environment. In China, most insulators are operated in moderately dirty or seriously dirty environments, and the order of importance of the above three elements is price, performance and maintenance costs with the cost and safety of the line and other factors taken into account.

It can be predicted that in the next few years, silicone rubber composite insulator will still be the first choice of the power sector due to its low price, excellent pollution resistance and simple maintenance.

4.2.3 Classification of Insulators

1. Rated voltage

According to the rated voltage, insulators are divided into: High-voltage insulator (above 500V), low-voltage insulator (below 500V). See Fig.4-8.

(a) High-voltage glass insulator   (b) Low-voltage ceramic insulator

Fig. 4-8 High-voltage Glass Insulator and Low-voltage Ceramic Insulator

2. Installation location

According to the installation location, insulators are divided into: Indoor type and outdoor type (see Fig.4-9).

(a) Indoor insulator (b) Outdoor insulator

Fig. 4-9　Indoor Insulator and Outdoor Insulator

3. Strluture form and use

According to the structure form and use, insulators are divided into: Post insulator, bushing insulator, cap and pin type suspension insulator.

(1) Post insulator.

Post insulator is a special type of insulating control that can play an important role in overhead transmission lines. In the early years, post insulators were mostly used for wire poles. As times goes by, there are many suspended insulators hanging from one end of the high-voltage wire connection tower to increase the creepage distance, which are usually made of silicone or ceramic. Insulators play two basic roles in overhead transmission lines, namely, supporting the wire and preventing the current from returning to ground. These two roles must be ensured. The insulators shall not fail by producing flashover breakdowns as a result of changes in the environment and electrical loading conditions, otherwise the insulators will become useless, which will damage the service and operating life of the entire line.

Post insulators conform to the provisions of *Technical Conditions of High-voltage Ceramic Post Insulators* (GB 8287.1), *Pollution-resistant Outdoor Bar-shaped Ceramic Post Insulators* (GB 12744), the international standard IEC *Testing of Indoor and Outdoor Ceramic or Glass Post Insulators for Systems with Nominal Voltages Greater than 1,000 Volts* and the IEC publication *Selection and Dimensioning of High Voltage for Polluted Conditions*, with following characteristics:

① High mechanical strength, small dispersion, safe and reliable operation.
② Good low temperature performance.
③ Good pollution resistance.
④ High earthquake resistance.
⑤ Low radio interference.

(2) Bushing insulator.

Bushing insulators are used to lead buses through walls or ceilings in the house and to lead them from the house to the outside, or to lead the current carrying part of an electrical equipment (e.g. circuit breaker, transformer, etc.) with an enclosed housing out of the housing. Bushing insulators are also known as wall-through bushings, or bushings for short.

Wall-through bushings can be divided into indoor type and outdoor type according to the installation location, and can be divided into conductor type and bus type according to the structure form. For conductor type bushing, the current carrying conductor and the insulating part are made into a whole, the conductor material is copper or aluminum, and the conductor section is rectangular or circular; The bus type bushing itself does not contain a current carrying conductor, and the current carrying bus is installed in the bushing.

(3) Cap and pin type suspension insulator (see Fig. 4-10).

Suspension insulators are mainly used in outdoor power distribution units and overhead lines of 35 kV and above, and metal fittings are generally composed of steel caps and pin-shaped steel legs. Therefore, cap and pin type suspension insulators are also known as cap-leg suspension insulator strings, which are divided into spherical and groove types according to the connection mode of caps and legs.

Suspension insulators are composed of insulating parts (ceramic or tempered glass), iron caps and iron legs. The pollution flashover voltage of bell hood shaped pollution proof insulators is 20%—50% higher than that of ordinary

1—Feet; 2—Hat.

Fig. 4-10 Cap and Pin Type Suspension Insulator

insulators; double-layer umbrella shaped pollution proof insulators have the advantages of large leakage distance, open umbrella, smooth skirt, low ash accumulation rate and good self-cleaning performance; straw hat-shaped pollution proof insulators have the advantages of low contamination rate and good self-cleaning performance.

In practice, suspension insulators form insulator strings according to the voltage of the units. The thick end of the leg of one insulator is led into the cap of the other insulator and locked with a special spring lock. The number of insulators per string is different at different voltage classes, generally not less than 3 pieces for 35 kV, not less than 7 pieces for 110 kV, not less than 13 pieces for 220 kV, not less than 19 pieces for 330 kV, and not less than 24 pieces for 500 kV. The number of insulators can be increased according to the degree of local pollution, and pollution proof suspension insulators shall be used for serious pollution.

4. The use

According to the use, insulators are divided into: Insulator for power station, insulator for electrical appliance, insulator for line.

(1) Insulator for power station is mainly used to support and secure the hard bus of

indoor and our door power distribution units of power plants and substations, and to insulate the bus from the ground. According to the different roles, the insulators for power station are divided into post insulators and bushing insulators.

(2) Insulator for electrical appliance is mainly used to secure the current carrying part of an electrical appliance. Insulators for electrical appliance are also divided into post insulators and bushing insulators. Post insulators are used to secure the current carrying part of an electrical appliance without an enclosed housing; bushing insulators are used to lead the current carrying part of an electrical appliance with an enclosed housing (e.g., circuit breaker, transformer, etc.) out of the housing.

(3) Insulator for line is mainly used to solidify overhead transmission and distribution wires and soft buses of outdoor power distribution units, and to insulate them from the grounding part. There are four types: Pin, suspension, butterfly and ceramic crossarm.

4.2.4 Insulator Fault and Preventive Measures

The insulator is one of the important components on the distribution line. It insulates the conductor from the ground, and once the insulator is damaged, the loss of insulation of the live wire to the ground will cause the grounded short circuit or the interphase short circuit fault, resulting in the interruption of the power supply of the line.

1. Fault of insulator damaged due to poor manufacturing

Good and qualified insulators have the characteristics of fine and white ceramics and uniform surface enamel, and can achieve the qualified standard in all kinds of electrical tests and temperature difference tests. Electrical tests include: Dry flashover test, wet flashover test, voltage withstand test, temperature difference test, etc. Insulators of poor quality cannot pass the above tests. In recent years, pin insulators and suspension insulators with poor ceramics and poor performance have been used on the distribution lines erected in rural areas. Accidents such as flashover, breakdown and burst occur shortly after the line is put into operation or after the rainy season.

A common characteristic of pin insulators and suspension insulators with poor ceramics is that they absorbs moisture with insulation rapidly reduced on rainy days, leading to flashover and breakdown. When the poor quality insulator is broken, it is found that its ceramic is rough and slightly tawny, and the ink spreads rapidly after dropping on it; sulfur and lead oxide are used as gluing agents to glue iron leg in some pin insulators, which are susceptible to thermal expansion and cause insulators to blow up.

Preventive measures: Before line erection, insulators shall be tested. When insulators produced by unknown manufacturers or non-electrical factories are used, comprehensive electrical tests shall be carried out, and dry flashover, wet flashover and voltage withstand tests shall be carried out after immersing the insulators for several hours. Insulators shall also be sampled for the temperature difference test, and the insulators failing the test after immersion in water can not be used on the line.

2. Fault of insulator damaged due to improper construction

In the process of erecting lines, due to carelessness in transportation or construction, insulators are subjected to damage, such as cracks, breakage, enamel detachment and so on. The insulators with a crack and mechanical damage area of more than 100 mm^2, if used on the line, will have breakdown or flashover faults in rainy season after being put into operation.

Preventive measures: The condition of insulators shall be carefully observed and checked before use to avoid using insulators with excessive crack and hard damage areas.

3. Insulator damaged due to external factors

Insulators on lines are sometimes cracked, damaged, broken, etc., when hit by thrown bricks, stones, or slingshot projectiles. After the insulator is cracked or severely damaged, it can sometimes continue to operate in sunny weather, but the insulation drops sharply in light rain and heavy fog weather, resulting in flashover or insulation breakdown and grounding fault.

Preventive measures: Defective insulator shall be replaced immediately when it is found, but if the insulator is lightly damaged and the damaged area is away from the wire, the damaged area may be coated with insulated paint and need not be replaced.

4. Fault of insulator caused by wearing out

Insulators in normal operation are subjected to high voltages and also to large tensile forces. After a long period of operation, they will be gradually wearing out, resulting in insulation degradation, so that fault occurs. The insulation resistance of insulators seriously worn out has been reduced to the extent that they can not continue to operate and need to be replaced immediately.

Preventive measures: The operator shall use the live test method. When the insulator worn out is found, it shall be removed in time. In addition, the insulation resistance of the insulator can be measured by using the insulation megger at the time of line power failure, and the lowest insulation electron of the qualified insulator in operation shall be more than 300 MΩ.

模块三　高压熔断器

一、熔断器的原理

熔断器是串联在电路中的一个最薄弱的导电环节，其金属熔体是一个易于熔断的导体。在正常工作情况下，由于通过熔体的电流较小，熔体的温度虽然上升，但不致达到熔点，熔体不会熔化，电路能可靠接通。一旦电路发生过负荷或短路故障时，电流增大，过负荷电流或短路电流对熔体加热，熔体由于自身温度超过熔点，在被保护设备的温度未达到破坏其绝缘之前熔化,将电路切断,从而使线路中的电气设备得到了保护。

熔断器的工作过程大致可分为以上四个阶段(见图 4-11),显然,熔断器的动作时间为上述四个过程所经过时间的总和。熔断器的开断能力决定于熄灭电弧能力的大小。熔体熔化时间的长短,取决于通过的电流的大小和熔体熔点的高低。当电路中通过很大的短路电流时,熔体将爆炸性地熔化并气化,迅速熔断;当通过不是很大的过电流时,熔体的温度上升得较慢,熔体熔化的时间也就较长。熔体材料的熔点高,则熔体熔化慢、熔断时间长;反之,熔断时间短。

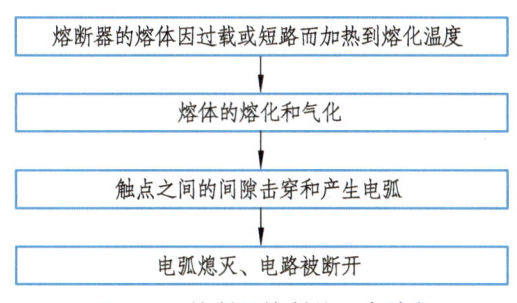

图 4-11 熔断器熔断的四个阶段

二、熔断器的结构

熔断器主要由金属熔断体、载熔件和底座组成。另外,有的熔断器还具有熔管、充填物、熔断指示器等结构部件。

(1)熔断体:熔断体是熔断器的主要部分,简称熔体。熔体是熔断器的核心部件,它是一个最薄弱的导电环节,正常工作时起导通电路的作用,在故障情况下熔体将首先熔化,从而切断电路实现对其他设备的保护。熔体可分为高熔点熔体和低熔点熔体。低熔点材料(如铅、锌、锡等)电阻率较大,所制成的熔体截面也较大,在熔化时将产生大量的金属蒸气,使电弧不易熄灭,所以这类熔体一般用于 500 V 及以下的低压熔断器中起过负荷保护;高熔点材料(如铜、银等)电阻率较低,所制成的熔体截面可较小,有利于电弧的熄灭,这类熔体一般用作短路保护。

高熔点材料应用在小而持续时间长的过负荷时,熔体不易熔断,结果使熔断器损坏。为此,在铜或银熔体的表面焊上小锡球或小铅球,当熔体发热到锡或铅的熔点时,锡或铅的小球先熔化,而渗入铜或银的内部,形成合金,电阻增大,发热加剧,同时熔点降低,首先在焊有小锡球或小铅球处熔断,形成电弧,从而使熔体沿全长熔化。这种方法称为"冶金效应"法,亦称金属熔剂法。

熔体以两个字母表示,如"gG""gM""aM"等。第一个字母"g"或"a"是按分断电流范围的不同来分,"g"熔体又称为全范围分断能力熔体,即在规定条件(包括电压、功率因数、时间常数等)下,能分断其分断能力范围内的所有电流;"a"熔体是部分范围分断能力熔体,在电路中作后备保护用,能分断四倍额定电流至额定分断电流之间的电流。第二个字母"G"或"M"是熔体按使用类别分,"G"类熔体为一般用途的熔体，"M"类熔体为电动机保护用熔体。

（2）载熔件：熔断器的可动部件，用于安装和拆御熔体。通过其接触部分将熔体固定在底座上，并将熔体与外部电路连接。载熔件通常采用触点的形式。

（3）熔断器底座：熔断器的固定部件，装有供电路连接的端子。包括绝缘件和其他必需的所有部件。绝缘件用于实现各导电部分的绝缘和固定。

（4）熔管：是熔断器的外壳，用于放置熔体，可以限制熔体电弧的燃烧范围，并具有一定的灭弧作用。

（5）充填物：一般采用固体石英砂，它是一种导热率很高的绝缘材料，用于冷却和熄灭电弧。石英砂填料之所以有助于灭弧，因为石英砂具有很大的热惯性与较高绝缘性能，并且因它为颗粒状，同电弧的接触面较大，能大量吸收电弧的能量，使电弧很快冷却，从而加快电弧熄灭过程。

（6）熔断指示器：用于反映熔体的状态，即完好或已熔断。

三、高压熔断器的分类及型号

在高压电网中，高压熔断器可作为配电变压器和配电线路的过负荷与短路保护，也可作为电压互感器的短路保护。

高压熔断器按照使用环境，分为户内式和户外式；按结构特点，分为支柱式和跌落式；按工作特性，可以分为限流型和非限流型。见图 4-12、4-13。

图 4-12 高压限流熔断器

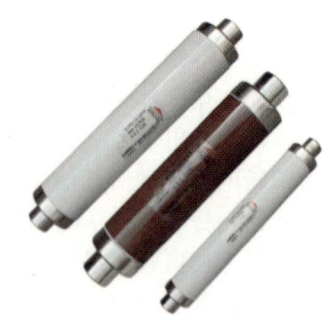

图 4-13 高压限流熔断器的连接方式

高压熔断器的型号含义：
国内型号：　　　　　　　　　　等同的国外型号：

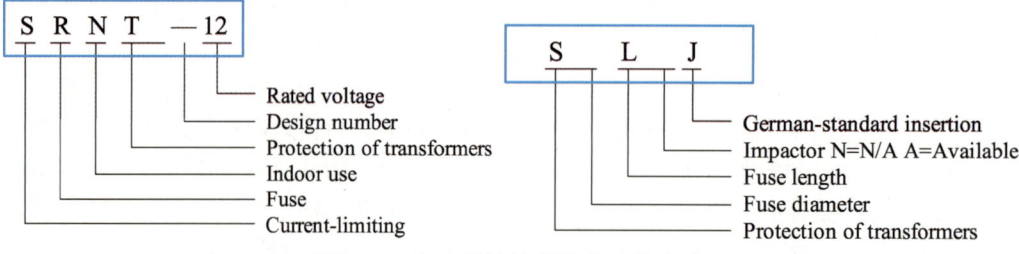

图 4-14 高压限流熔断器的连接方式

（1）户内高压熔断器。

户内高压熔断器全部是限流型熔断器,其熔体装在充满石英砂的密封瓷管内,当短路电流通过熔件使其熔断时,电弧产生在石英砂的填料中,受到石英砂颗粒间狭沟的限制,弧柱直径很小,同时电弧还受到很多的气体压力作用和石英砂对它的强烈冷却,所以限流式熔断器灭弧能力强,在短路电流未达到最大值时,就将电弧很快熄灭,因而可限制短路电流的发展,大大减轻了电气设备所受危害的程度,降低了对被保护设备动、热稳定性的要求。因它在开断电路时无游离气体排出,所以在户内配电装置中广泛采用。

户内高压熔断器主要由 RN1 和 RN2 型两种。RN1 型熔断器适用于 3~35 kV 的电力线路和电力变压器过载和短路保护;RN2 型专门用于 3~35 kV 电压互感器的短路保护。二者的结构基本相同。

RN1 型熔断器外形如图 4-15 所示,它由瓷质熔管、触座、支柱绝缘子及底座组成。RN2 型熔断器的熔体由三种不同截面的铜丝连接而成,绕在陶瓷芯上,但无指示器,在运行中,当高压熔体熔断时,根据声光信号及电压互感器二次电路中仪表指示的消失来判断。

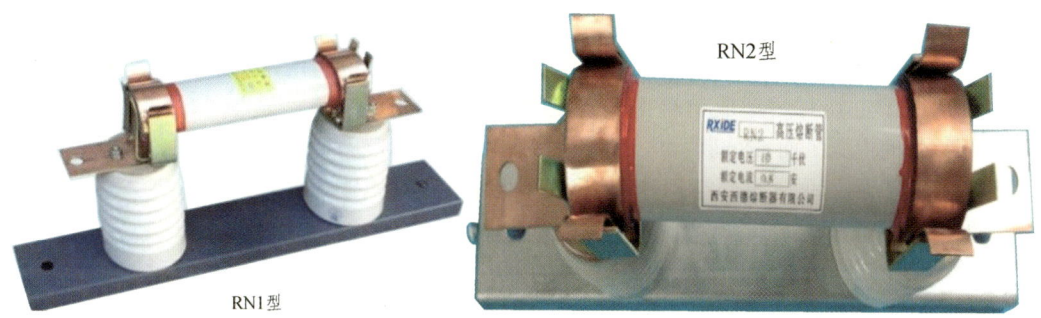

图 4-15　RN1 型和 RN2 型熔断器外观

（2）户外高压熔断器。

户外式高压熔断器主要用于输电线路和电力变压器的过负荷与短路保护。户外式高压熔断器型号较多,按其结构和工作原理可分跌落式熔断器和支柱式熔断器。

户外跌落式熔断器（见图 4-16）具有经济实惠、操作方便、适应户外环境性强等特点,广泛应用于 10 kV 架空配电线路的支线及用户进线处、35 kVA 以下容量的配电变压器一次侧以及电力电容器等设备作为过载或短路保护和进行系统、设备投、切操作之用。它安装在 10 kV 配电线路分支线上,可缩小停电范围,因其熔断时有一个明显的断开点,为检修线路和设备创造了一个安全作业环境,增加了检修人员的安全感。户外跌落式熔断器主要由熔丝具、熔丝管和熔丝元件三部分构成。在熔丝管内装有用桑皮纸或钢纸等制成的消弧管。熔管两端的上动触头和下动触头依靠熔断体系紧,将上动触头推入鸭嘴凸出部分后,磷铜片等制成的上静触头顶着上动触头,故而将熔管牢固地卡在鸭嘴里。当短路电流通过电路使熔体熔断时,将产生电弧,管内衬的钢纸管在电弧作用下产生大量气体,在电流过零时将电弧熄灭。由于熔体熔断,在熔管的上下动触头弹簧片的作用下，熔管迅速跌落，使电路断开，切除故障段线路或者故障设备。

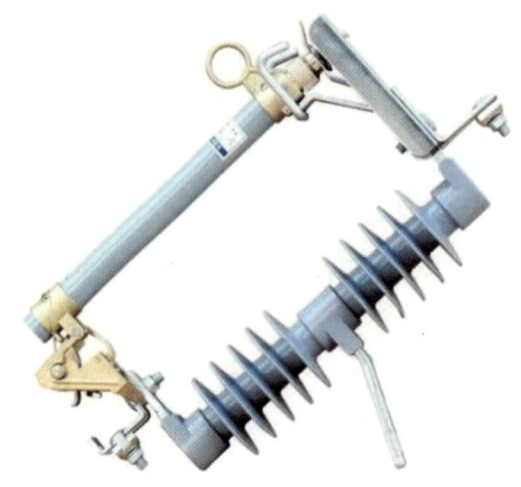

图 4-16　户外跌落式熔断器

这种熔断器是靠熔管产气吹弧和迅速拉长电弧而熄灭，它还采用了"逐级排气"的新结构。图中熔管上端有管帽，在正常运行时是封闭的，可防雨水滴入。分断小的故障电流时，由于上端封闭形成单端排气（纵吹），使管内保持较大压力，有利于熄灭小故障电流产生的电弧；而在分断大电流时，由于电弧使消弧管产生大量气体，气压增加快，上端管帽被冲开，而形成两端排气，以免造成熔断器机械破坏，有效地解决了自产气电器分断大、小电流的矛盾。

RXW-35 型限流式熔断器主要用于保护电压互感器，结构如图 4-17 所示。熔断器由瓷套、熔管及棒形支持绝缘子和接线端帽等组成。熔管装于瓷套中，熔件放在充满石英砂填粒的熔管内。熔断器的灭弧原理与 RN 系列限流式有填料高压熔断器的灭弧原理基本相同，均有限流作用。

图 4-17　RXW-35 系列流熔断器外型图

为了保证在暂时性故障后迅速恢复供电,有些高压熔断器具有单次重合功能。例如,RW3-10Z 型单次重合熔断器,它具有两根熔件管,平时只有一根接通工作,当这根熔件管断开后,相隔一定时间(约 0.3 s 以内),另一根熔件管借助重合机构而自动重合,恢复供电。

Module 3　High-voltage Fuse

4.3.1　Principle of the Fuse

A fuse is one of the weakest conductive links in a series circuit, and its metal fuse link is an easily fusible conductor. Under normal operating conditions, due to the small current through the fuse link, the temperature of the fuse link rises, but does not reach the melting point, the fuse link will not melt, so the circuit can be reliably connected. Once the circuit overload or short-circuit fault occurs, the current increases, the overload current or short-circuit current heats the fuse link. Due to its own temperature exceeding the melting point, the fuse link melts before the temperature of the protected equipment reaches a level that destroys the insulation, cutting off the circuit, thus allowing the electrical equipment in the line to be protected.

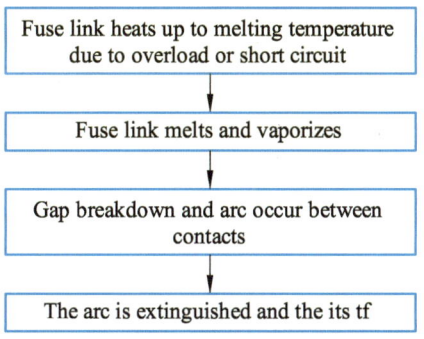

Fig. 4-11　Four Stages of Fuse Fusing

The working process of the fuse can be roughly divided into the above four stages (see Fig.4-11). Obviously, the action time of the fuse is the sum of the time of the above four stages. The cut-off capacity of a fuse is determined by the capacity of extinguishing arc. The melting duration of the fuse link depends on the magnitude of current through it and the melting point of the fuse link. When a large short-circuit current is passed through the circuit, the fuse link will melt explosively and vaporize, rapidly fusing; when a not very large overcurrent is passed through the circuit, the temperature of the fuse link rises slowly and the fuse link takes a long time to melt. If the melting point of the fuse link material is high, the fuse link melts slowly and takes a long time to fuse; otherwise, the fusing time is short.

4.3.2 Structure of the Fuse

The fuse is mainly composed of a metal fuse link, a fuse carrier and a base. In addition, there are fuses that have structural components such as a cartridge, a filling, and a fusing indicator.

(1) Fuse link. It is the main part of the fuse. Fuse link is the core component of the fuse, it is the weakest conductive link, and it can be used for circuit conduction in the normal operation. In the case of fault, the fuse link will melt first, so as to cut off the circuit to protect other equipment. Fuse links can be divided into fuse links with high melting point and fuse links with low melting point. Materials with low melting point (such as lead, zinc, tin, etc.) have high resistivity. The section of the fuse link made of such materials is large, and a large amount of metal vapor will be generated during melting of the fuse link, making the arc not easy to extinguish. Therefore, such kind of fuse link is generally used for overload protection in low-voltage fuses up to 500 V. Materials with high melting point (such as copper, silver, etc.) have high resistivity. The section of the fuse link made of such materials is small, conducive to the arc extinguishing. Therefore, such kind of fuse link is generally used for short-circuit protection.

When materials with high melting point are applied to small but long lasting overloads, the fuse link does not fuse easily and as a result the fuse is damaged. To this end, in case the surface of the copper or silver fuse link is welded with a small tin ball or small lead ball, when the fuse link heats up to the melting point of tin or lead, the small tin or lead ball melts first, and the melting liquid penetrates into the interior of copper or silver, forming an alloy, so the resistance increases, the heat intensifies, and the melting point decreases, and the place welded with small tin ball or small lead ball is first fused, forming an arc, so that the fuse link melts along the full length. This method is called the "metallurgical effect" method, also known as the metal flux method.

The fuse link is represented by two letters, such as "gG", "gM", "aM", etc. The first letter "g" or "a" depends on the different range of breaking current. "g" fuse link is also known as the fuse link for the full range of breaking capacity. To be specific, in the specified conditions (including voltage. power factor, time constant, etc.), it can break all the current within its breaking capacity; "a" fuse link is also known as the fuse link for the partial range of breaking capacity, used for backup protection in circuits, capable of breaking currents between four times the rated current and the rated breaking current. The second letter "G" or "M" is the fuse link according to the use of the category. "G" fuse link is for general purposes, and "M" fuse link is for motor protection.

(2) Fuse carrier. It is a movable component of a fuse for mounting and dismounting the

fuse. The fuse carrier can secure the fuse to the base and connects the fuse to the external circuit by means of its contact part. Fuse carriers are usually in the form of contacts.

(3) Fuse base. It is a fixed component of the fuse fitted with terminals for supply circuit connection. It includes insulating parts and all other necessary components. Insulating parts are used to insulate and secure the conductive parts.

(4) Cartridge. It is the housing of the fuse used to place the fuse link, can limit the combustion range of the fuse link arc, and has a certain arc extinguishing effect.

(5) Filling. Solid quartz sand is generally adopted as the filling, which is an insulating material with high thermal conductivity, and is generally used to cool and extinguish arc. The reason why the quartz sand filling helps to extinguish the arc is that the quartz sand has a large thermal inertia and high insulating properties, and it is granular, the contact surface with the arc is large, it can absorb a lot of energy of the arc, so that the arc cools quickly, thereby speeding up the arc extinguishing process.

(6) Fusing indicator. It is used to reflect the state of the fuse link, i.e. intact or fused.

4.3.3 Classification and Model of High-voltage Fuse

In high-voltage power grid, high-voltage fuses can be used for overload and short-circuit protection for distribution transformers and distribution lines, as well as short-circuit protection for voltage transformers.

High-voltage fuses are divided into indoor and outdoor types according to the use environment; high-voltage fuses are divided into post type and drop-out type according to the structural characteristics; high-voltage fuses are divided into current-limiting type and non-current-limiting type according to the operating characteristics. See Fig. 4-12 and 4-13.

 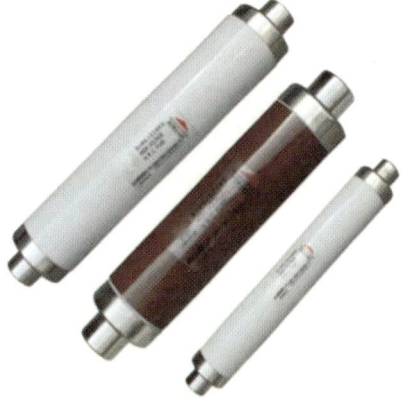

Fig. 4-12 High-voltage Current-limiting Fuse Connection   Fig. 4-13 High-voltage Current-limiting Fuse

Meaning of the model of the high-voltage fuse:

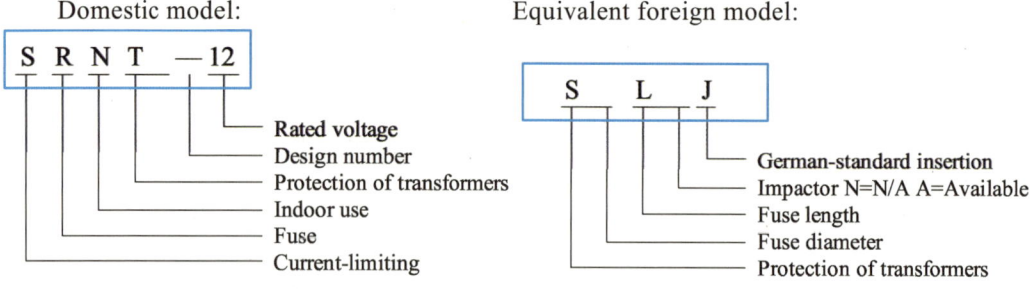

Fig. 4-14 High-voltage Fuse Connecting Ways

(1) Indoor high-voltage fuse.

Indoor high-voltage fuse is current-limiting fuse, and the fuse link is installed in a enclosed ceramic cartridge filled with quartz sand. When the short-circuit current passes through the fuse link to make it fuse, the arc is generated in the quartz sand filling and limited by the narrow groove between the quartz sand particles, the diameter of the arc column is very small, and the arc is also subjected to a lot of gas pressure and the strong cooling from quartz sand, so the current-limiting fuse has a strong arc extinguishing ability, and the arc is quickly extinguished when the short-circuit current does not reach the maximum. Therefore, it can limit the development of short-circuit current, greatly reduce the harm to electrical equipment, and reduce the requirements for the dynamic and thermal stability of the protected equipment. It is widely used in indoor power distribution units because it has no ionized gas discharge hen the circuit is cut off.

Indoor high-voltage fuses mainly include RN1 type and RN2 type. RN1 type fuses are suitable for overload and short-circuit protection of power lines and power transformers from 3—35 kV; RN2 type fuses are specially designed for short-circuit protection of 3—35 kV voltage transformers. The structure of both is basically the same.

The shape of RN1 type fuse is shown in Fig. 4-15. It consists of ceramic cartridge, contact base, post insulator and base. The fuse link of RN2 type fuse is connected by three kinds of copper wires with different sections, wound on the ceramic core, but without indicator. In operation, when the high-voltage fuse link fuses, it is determined according to the acoustic and optical signals and the disappearance of the meter indication in the secondary circuit of the voltage transformer.

Fig. 4-15 Appearance of RN1 Type and RN2 Type Fuses

(2) Outdoor high-voltage fuse.

Outdoor high-voltage fuses are mainly used for overload and short-circuit protection of transmission lines and power transformers. There are many types of outdoor high-voltage fuses, which can be divided into drop-out fuses and post fuses according to their structure and working principle.

Outdoor drop-out fuse (see Fig. 4-16) is economical, easy to operate, and adaptable to outdoor environment. It is widely used in the branch lines and user's incoming line of 10 kV overhead distribution lines, the primary side of distribution transformers with a capacity below 35 kVA, and power capacitors and other equipment for overload or short-circuit protection and the switching operation of systems and equipment. It is installed in the branch lines of 10 kV distribution lines, and can reduce the scope of power failure. Because it has an obvious fused point, it creates a safe operating environment for maintaining lines and equipment, and increases the safety of the maintainers. Outdoor drop-out fuse is mainly composed of three parts: Fuse fixture, cartridge and fuse element. An arc extinguishing tube made of mulberry paper or vulcanized fibre paper, etc., is provided inside the cartridge. The upper and lower moving contact terminals at both ends of the cartridge are secured by the fuse link. After the upper moving contact terminal is pushed into the protruding part of the duckbilled part, the upper static contact terminal made of phosphorus and copper sheet is pressed against the upper moving contact terminal, so the cartridge is firmly stuck in the duckbilled part. When the short-circuit current passes through the circuit to fuse the the fuse link, an arc will be produced. A large amount of gas will be produced in the vulcanized fibre paper tube lined in the cartridge under the action of the arc, and the arc will be extinguished at zero crossing of current. As the fuse link fuses, under the action of the spring plates of the upper and lower moving contact terminals of the cartridge, the cartridge drops down quickly, breaking the circuit and cutting of the faulty section of the line or the faulty equipment.

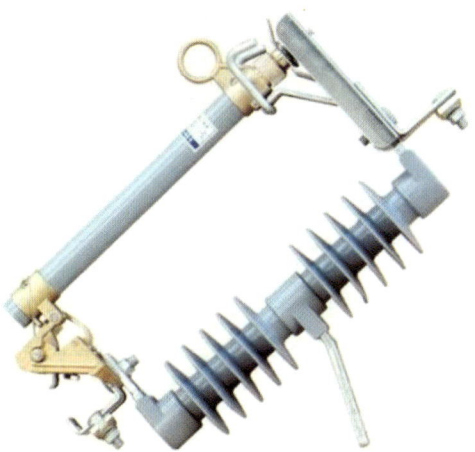

Fig. 4-16　Outdoor Drop-out Fuse

For this type of fuse, gas is produced in the cartridge to blow the arc and rapidly stretch the arc to extinguish it. Moreover, a new structure of "exhaust step by step" is adopted. The cartridge shown has a cap on the top end, which is closed during normal operation to prevent rainwater from dripping in. At the time of breaking off the small fault current, the exhaust (longitudinal blowing) at the single end is allowed because the top end is closed, which keeps a large pressure in the cartridge, which is beneficial to extinguish the arc caused by the small fault current. However, at the time of breaking off the large current, because a large amount of gas is produced in the arc extinguishing tube as a result of arc, the air pressure increases faster, the cap on the top end is blown away, and the exhaust at both ends is allowed, so as to avoid mechanical damage to the fuse, which effectively solves the contradiction of breaking off large and small current in self-produced gas appliances.

RXW-35 type current-limiting fuse is mainly used to protect the voltage transformer, with the structure shown in Fig. 4-17. The fuse consists of a ceramic bushing, a cartridge and a bar-shaped support insulator and terminal caps. The cartridge is installed in a ceramic bushing, and the fuse link is placed inside the cartridge filled with quartz sand. The fuse and the RN series of current-limiting stuffed high-voltage fuse have basically same arc extinguishing principle, both of which have current-limiting effect.

Fig. 4-17　Appearance of RXW-35 Series Current-limiting Fuse

To ensure rapid restoration of power after a temporary fault, some high-voltage fuses have a single reclosing function. For example, RW3-10Z type single reclosing fuse has two cartridges. Usually only one cartridge is connected to work. When the cartridge is disconnected, after a certain period of time (about 0.3s), another cartridge is automatically reclosed to restore power by virtue of the reclosing mechanism.

模块四 补偿设备

一、补偿设备的作用和地位

在电力系统中,补偿设备指的是无功补偿设备,在电力系统中主要起到提高电路功率因数的作用,能够降低供电变压器及输电线路的功率损耗,提高供电效率,改善供电环境。

无功补偿设备是电力系统中必不可少的重要装置,合理使用无功补偿设备可以最大限度减少电网的损耗,提高供电质量。反之,如果无功补偿设备使用不当,可能造成供电系统电压波动,谐波增大等问题。

二、补偿设备的基本原理

电网输出的功率分为两部分,一是有功功率:直接消耗电能,将电能转化为机械能、热能、化学能或声能,再利用这些能量做功,这部分功率就称为有功功率。二是无功功率:将电能转化为另一种形式的能(如建立电场需要的电场能,建立磁场需要的磁场能),这种能是电气设备能够做功的基本条件,并且这种能量在电网中与电能进行周期性的转换,这部分功率称为无功功率。无功补偿相量图见图 4-18。

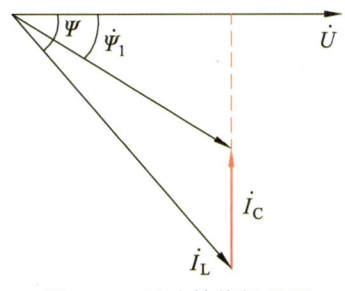

图 4-18 无功补偿相量图

补偿前,正弦交流电路从电源处获得的有功功率为 $P=UI_L\cos\psi$,无功功率为 $Q=UI_L\sin\Psi$;对于感性负载,在负载两端并联电容后,供电线路上多了一部分电容电流 $\dot{I}_c$,则补偿后,正弦交流电路从电源处获得的有功功率为 $P=UI\cos\Psi'$,无功功率为 $\theta'=UI\sin\Psi'$。

由相量图可知,补偿以后有功功率不变,无功功率减小,通过线路的电流减小,导线电阻的能量损耗和压降减小,因此能使电源设备的容量得到合理利用。

三、常用的补偿方式

(1)集中补偿:在低压配电线路中集中安装并联补偿电容器组(见图 4-19)。
(2)分组补偿:在配电变压器低压侧和用户车间配电屏安装并联补偿电容器。
(3)单台电动机就地补偿:在单台电动机处安装并联电容器等。

图 4-19 配电线路并联电容器组

四、无功补偿设备

无功补偿设备大致可以分为三类：调相机（见图 4-20）、静止无功补偿设备（Static Var Compensator，SVC）、静止无功发生装置（Static Var Generator，SVG）。

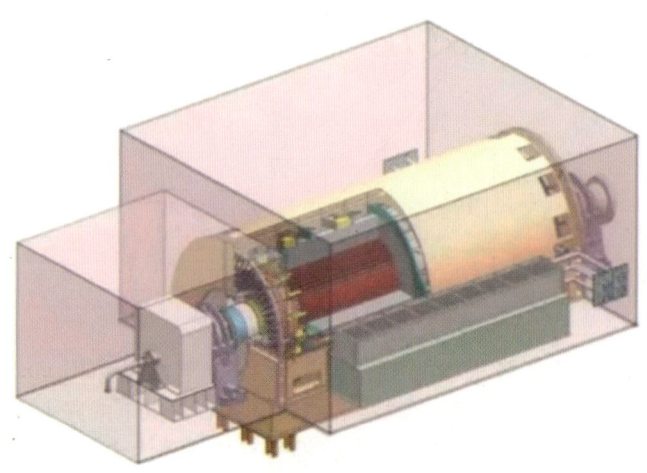

图 4-20 调相机

调相机也称同步调相机、同步补偿机，是出现较早的一类无功补偿设备。调相机实际是一台空载运行的同步电动机，利用同步电动机在不同励磁电流下的发出或吸收无功电流的能力起到无功补偿作用。当正常励磁时，调相机的电枢电流接近于零；过励磁时，调相机向电网发出无功电流；欠励磁时，调相机从电网中吸收无功电流。为方便运行起见，调相机一般与发电厂中的同步发电机组或负荷端的异步电动机组安装在一起，容量较大的调相机还需要采用氢气冷却。以上缺点均大大限制了调相机的应用范围，目前除在高压直流输电线路的终端作为动态无功支持外，已很少使用。

SVC 是目前应用最为广泛的一类无功补偿设备。但就字面而言，SVC 中的"Static"即静止，是相对于调相机的旋转而言，因此除了调相机和 SVG 之外，凡是用电感或电容进行无功补偿的装置均可称作 SVC。按国际大电网会议的定义，SVC 可分为以下 7 类：

（1）机械投切电容器（MSC）；

（2）机械投切电抗器（MSR）；

（3）自饱和电抗器（SR）；

（4）晶闸管控制电抗器（TCR）；

（5）晶闸管投切电抗器（TSC）；

（6）晶闸管投切电抗器（TSR）；

（7）自换向或电网换向转换器（SCC/LCC）。

实际上以上 7 类仍未能涵盖全部 SVC 设备，一般应慎重使用 SVC 这一名词，因为其所能涵盖的范围过于宽泛。

在种类繁多的 SVC 设备中，一般可按控制/投切设备的种类分为机械投切型及电力电子型两大类，通常所称的 SVC 设备也指这两类。前者一般包括机械投切电容器（MSC）、机械投切电抗器（MSR）等，共同特点是采用机械投切开关如接触器、遥控断路器等作为投切设备，其优点是鲁棒性较好、不易受谐波干扰等，缺点则是响应时间长、一般只能分级投入不易实现动态无功补偿等。电力电子型无功补偿装置一般包括晶闸管控制电抗器（TCR）、晶闸管投切电容器（TSC）、晶闸管投切电抗器（TSR）等，其优点是响应速度快、控制精度较好、能实现过零投切、具备动态补偿能力等，缺点则是电力电子器件易受谐波干扰；随着电力电子技术的发展，电力电子型 SVC 设备有逐渐取代机械投切型 SVC 设备的趋势。

SVG 与 SVC 一样，也是一种"静止"设备，是近年来出现的新一代无功补偿设备。SVG 以 IGBT 模块构成的电压源型逆变器为核心，通过实时检测负载电流波形，通过上下调节，可以吸收或者发出满足要求的无功电流，实现动态无功补偿。SVG 具有响应速度快、吸收无功连续、产生的高次谐波量小、调节范围广、损耗与噪音小等突出优点。但 SVG 的主要问题是电压等级和容量受 IGBT 器件性能限制，较难做到高电压等级和较大规模容量，目前一般在 1kV 以下电压等级应用较为广泛。

Module 4　Compensator

4.4.1　Role and Status of Compensator

In the power system, compensator refers to var compensator, and is mainly used to improve the power factor of the circuit in the power system, which can reduce the power loss of the power supply transformers and transmission lines, increase the efficiency of power supply and improve the power supply environment.

Var compensator is an essential and important device in the power system, and the loss of the power grid can be minimized and the quality of power supply can be improved provided that var compensator is reasonably used. If var compensator is not used properly, it may cause voltage fluctuations in the power supply system and increase in harmonics.

4.4.2 Fundamentals of Compensator

The power output of the power grid is divided into two parts. One part is active power: Electric energy is consumed directly and is converted into mechanical energy, thermal energy, chemical energy or sound energy, which is then used to do work. This part of power is called active power. The other part is reactive power: The electric energy is converted into another form of energy (such as the electric field energy needed to establish the electric field and the magnetic field energy needed to establish the magnetic field), which is the basic condition for electrical equipment to do work, and is periodically converted with electric energy in the power grid. This part of power is called reactive power (see Fig. 4-18).

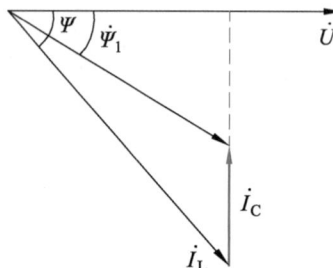

Fig. 4-18 Phasor Diagram of Reactive Power Compensation

Before compensation, the active power obtained by the sinusoidal AC circuit from the power supply is: $P=UI_L\cos\psi$ and the reactive power is: $Q=UI_L\sin\psi$. For inductive loads, after the shunt capacitor at both ends of the load, a part of the capacitive current $\dot{I}_c$ is added to the power supply line, and then after compensation, the active power obtained by the sinusoidal AC circuit from the power supply is: $P=UI\cos\psi'$ and reactive power is: $Q'=UI\sin\psi'$.

As can be seen from the phasor diagram, after compensation, the active power remains unchanged, the reactive power decreases, the current through the line decreases, and the energy loss and voltage drop of the wire resistance decreases, so the capacity of the power supply equipment can be reasonably utilized.

4.4.3 Commonly Used Compensation Methods

(1) Centralized compensation: Centralized installation of shunt compensation capacitor banks in low-voltage distribution lines (see Fig. 4-19).

(2) Compensation in banks: Shunt compensation capacitors are installed on the LV side of the distribution transformer and in the distribution panel of the user's workshop.

Fig. 4-19　Shunt Capacitor Bank of Distribution Line

(3) Local compensation for a single motor: Shunt capacitors are installed at a single motor.

4.4.4　Var Compensator

Var compensators can be broadly categorized into three types: Phase modifier, static var compensator (SVC), and static var generator (SVG).

The phase modifier (see Fig. 4-20) also refers to the synchronous phase modifier and the synchronous compensator, which is the earlier class of var compensator. The phase modifier is actually a no-load synchronous motor. The ability of the synchronous motor is used to emit or absorb reactive current under different exciting currents for reactive power compensation. When normally excited, the armature current of the phase modifier is close to zero; when overexcited, the phase modifier sends reactive current to the power grid; when under-excited, the phase modifier absorbs reactive current from the power grid. For ease of operation, the phase modifier is generally installed with synchronous generator units or load-side asynchronous motor units in the power plant, and the large-capacity phase modifier also requires hydrogen cooling. All of the above disadvantages greatly limit the scope of application of the phase modifier, and is rarely used except as dynamic reactive power support at the terminals of high-voltage DC transmission lines.

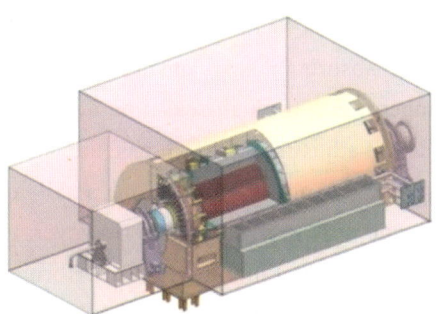

Fig. 4-20　Phase Modifier

SVC is currently the most widely used class of var compensator. However, literally, the "Static" (in SVC) state is relative to the rotation of the phase modifier, so except the phase modifier and SVG, any inductive or capacitive var compensator can be called SVC. As defined by the CIGRE, SVCs can be categorized into the following seven types:

(1) Mechanically switched capacitor (MSC);

(2) mechanically switched reactor (MSR);

(3) self-saturating reactor (SR);

(4) thyristor controlled reactor (TCR);

(5) thyristor switched capacitor (TSC);

(6) thyristor switched reactor (TSR);

(7) self-commutated converter or line-commutated converter (SCC/LCC).

In fact, the above seven types still do not cover all SVCs, and the term SVC shall generally be used with caution, as the scope it can cover is too broad.

The various SVCs generally can be divided into two categories, i.e. mechanically switched type and power electronic type according to the category of controlled/switched equipment. The SVCs, as commonly referred to, refer to the two categories. The former category generally includes mechanically switched capacitors (MSC), mechanically switched reactors (MSR), etc. The common feature is the use of mechanically switched switch such as contactor, remote-controlled circuit breaker, etc. as switched equipment, which has the advantages of favorable robustness, being less susceptible to harmonic interference, etc., and the disadvantages of long response time, generally only graded input allowed and difficulty in realizing dynamic reactive power compensation, etc. Power electronic var compensators generally include thyristor controlled reactor (TCR), thyristor switched capacitor (TSC), thyristor switched reactor (TSR) and so on, which have the advantages of fast response speed, favorable control accuracy, availability of the zero-crossing switching, and the ability to dynamically compensate and the disadvantages of the power electronics susceptible to harmonic interference; with the development of power electronic technology, power electronic SVCs have gradually replaced the mechanically switched SVCs.

SVG, like SVC, is a kind of "static" equipment, which is a new generation of var compensator emerging in recent years. With the voltage source inverter composed of IGBT module as the core, SVG can absorb or send out adequate reactive current, and achieve dynamic reactive power compensation through real-time detection of load current waveform and by means of up and down adjustment. SVG has the outstanding advantages of fast response speed, continuous reactive power absorption, small amount of generated higher harmonics, wide regulation range, low loss and noise. However, the main problem of SVG is that the voltage class and capacity are limited by the performance of IGBT devices, it is difficult to achieve high voltage class and large capacity, and it is generally widely used in the voltage class below 1 kV.

参考文献

[1] 高建. 电气设备检修[M]. 成都：成都时代出版社，2019.
[2] 杨迪. 变电检修技能培训教材[M]. 北京：中国电力出版社，2019.
[3] 姜聿涵. 变压器检修技能培训教材[M]. 北京：中国电力出版社，2019.
[4] 姜聿涵，杨冰."一带一路"变电设备检修专业培训教材　变压器检修（结构及附件篇）（中英文对照）[M]. 北京：中国电力出版社，2021.
[5] 华章，雷春."一带一路"变电设备检修专业培训教材　变压器检修（电气试验篇）（中英文对照）[M]. 北京：中国电力出版社，2021.
[6] 邓常飞，祝捷."一带一路"变电设备检修专业培训教材　变电检修（故障处理篇）（中英文对照）[M]. 北京：中国电力出版社，2021.